Internet of Nano-Things and Wireless Body Area Networks (WBAN)

Internet of Nano-Things and Wireless Body Area Networks (WBAN)

Fadi Al-Turjman

CRC Press
Taylor & Francis Group
Boca Raton London New York

CRC Press is an imprint of the
Taylor & Francis Group, an **informa** business
AN AUERBACH BOOK

CRC Press
Taylor & Francis Group
6000 Broken Sound Parkway NW, Suite 300
Boca Raton, FL 33487-2742

First issued in paperback 2022

© 2019 by Taylor & Francis Group, LLC
CRC Press is an imprint of Taylor & Francis Group, an Informa business

No claim to original U.S. Government works

ISBN 13: 978-0-367-19852-7 (hbk)
ISBN 13: 978-1-03-240148-5 (pbk)

DOI: 10.1201/9780429243707

Library of Congress Cataloging-in-Publication Data

Names: Al-Turjman, Fadi, author.
Title: Internet of nano-things and wireless body area networks (WBAN) / Fadi Al-Turjman.
Description: Boca Raton, FL : CRC Press/Taylor & Francis Group, 2019. |
Includes bibliographical references and index.
Identifiers: LCCN 2019008529 (print) | LCCN 2019011375 (ebook) | ISBN 9780429243707
(e) | ISBN 9780367198527 (hb : acid-free paper)
Subjects: LCSH: Body area networks (Electronics)
Classification: LCC TK5103.35 (ebook) | LCC TK5103.35 .A43 2019 (print) | DDC
004.67/8—dc23
LC record available at https://lccn.loc.gov/2019008529

Visit the Taylor & Francis Web site at
http://www.taylorandfrancis.com

and the CRC Press Web site at
http://www.crcpress.com

To my dearest parents, my brother, and my sisters.

To my wonderful wife and my little stars.

To my father..

أبي يا أبي....يا قطرة الندى يا شذى الياسمين المعطر

يا نفحة الحب....يا درة الشرق المؤصل

هل رأى الغرب عينا أبي؟

عينا أبي خبر من جنة الفردوس.. وشهب لم تزل بالافق الارحب

عينا أبي كواكب شهل.. يرسو بها سحر الشرق والليلك

اه ياابي .. بعينيك كم تمنيت لو أصلب

آه لو تعلمون من أي عطر أبي خلق ..

ابي انا.. نبراس علم.. جاء ليتم مكارم الأخلاق...و يا له من نبي

يا كرم زيتون.. يا غابات لوز.. يا ملجأ حب يا أبي

آه يا أبي....يا عود طيب عاش ليحرق

تمشي وتمشي وراءك تاريخ فخر حتى تخلد

يا من توالد من ثغره أحاديث الورود والطيب والزنبق

تميل على صدره المقدس.. لنشكو هموما ما كانت لتنتهي

أه يا أبي .. يا من سكن بأحداق عينيي.. ونبرة صوتي... حتى ظننت أني ابي

ولكن... اين أنا منك يا أبي

لولاك ما اضاءت سماء بأنجم ..

لولاك .. ولولا بريق بعينيك اي ليل ينجلي ..

لولاك يا أبي... أي شعر يكتب..

Contents

Preface ix

About the Author xi

Contributors xiii

1   Introduction 1
FADI AL-TURJMAN

2   Internet of Nano-Things (IoNT) & WBAN 5
FADI AL-TURJMAN, ENVER EVER,
AND HADI ZAHMATKESH

3   Nanosensors for the Internet of Nano-Things
(IoNT): An Overview 21
SEDA DEMIREL TOPEL AND FADI AL-TURJMAN

4   mm-Waves in the Internet of Nano-Things 45
FADI AL-TURJMAN AND JEHAD HAMAMREH

5   A Rational Routing Protocol for WBAN 59
FADI AL-TURJMAN

6   A Value-Based Caching Approach for WBAN 87
FADI AL-TURJMAN

7   Adaptive WBAN in the IoNT 113
FADI AL-TURJMAN AND HAMED OSOULI TABRIZI

8   A Cognitive Routing Protocol for WBAN 129
FADI AL-TURJMAN

9   Energy-Aware Routing Protocol for Nanosensor
Networks 151
FADI AL-TURJMAN AND KEMAL IHSAN KILIC

10 LCPC Code for Wireless Body Area Networks 177
SALAH A. ALABADY AND FADI AL-TURJMAN

11 Energy-Harvesting Methods for WBAN Applications 203
SÜLEYMAN MAHIRCAN DEMIR, FADI AL-TURJMAN,
AND ALI MUHTAROĞLU

Index 233

Preface

The Internet of Nano-Things (IoNT) is a system of nanoconnected computing devices, objects, and/or people that are provided with unique identifiers to transfer data over a computer/cellular network wirelessly. Thus, routing protocols and communication issues are considered among the most significant topics in the IoNT paradigm. The implementation of wireless body area networks (WBANs) is at the heart of this paradigm, and its development is a key issue in next generation IoNT. This book is dedicated to address major design aspects and challenges in realizing the nanosensing platforms in critical cloud and IoNT applications. Challenges vary from manufacturing at the nanoscale to reliability issues in wireless millimeter Waves (mm-Waves). The aim of this book is hence to focus on both the design and implementation aspects of the physical nanosensor and its wireless communication aspects in IoNT projects that are enabled by the WBAN. It is mainly focused on energy-efficient data gathering/delivery approaches and reliable IoNT systems.

By Fadi Al-Turjman

About the Author

Professor Fadi Al-Turjman received his Ph.D. in computer science from Queen's University, Kingston, Ontario, Canada, in 2011. He is a professor at Antalya Bilim University, Antalya, Turkey. Professor Al-Turjman is a leading authority in the areas of smart/cognitive, wireless, and mobile networks' architectures, protocols, deployments, and performance evaluation. His publication history spans over 200 publications in journals, conferences, patents, books, and book chapters, in addition to numerous keynotes and plenary talks at flagship venues. He has written and edited more than fifteen books about cognition, security, and wireless sensor networks' deployments in smart environments, published by Taylor & Francis Group and Springer. He has received several recognitions and best papers' awards at top international conferences. He also received the prestigious *Best Research Paper Award* from Elsevier Computer Communications Journal for the period 2015-2018, in addition to the *Top Researcher Award* for 2018 at Antalya Bilim University, Turkey. Professor Al-Turjman has led a number of international symposia and workshops in flagship communication society conferences. Currently, he serves as the lead guest editor for several well reputed journals, including the Elsevier *Computer Communications* and the IET *Wireless Sensor Systems*.

Contributors

Salah A. Alabady
Computer Engineering Department
University of Mosul
Mosul, Iraq

Fadi Al-Turjman
Department of Computer Engineering
Antalya Bilim University
Antalya, Turkey

Süleyman Mahircan Demir
Department of Electrical and
    Electronics Engineering
Middle East Technical University
    Northern Cyprus Campus
Kalkanli, Güzelyurt, Turkey

Enver Ever
Department of Computer Engineering
Antalya Bilim University
Antalya, Turkey

Jehad Hamamreh
Antalya Bilim University
Antalya, Turkey

Kemal Ihsan Kilic
Department of Computer Engineering
Middle East Technical University
    Northern Cyprus Campus
Kalkanli, Güzelyurt, Turkey

Ali Muhtaroğlu
Center for Sustainability
Middle East Technical University
    Northern Cyprus Campus
Kalkanli, Güzelyurt, Turkey

Hamed Osouli Tabrizi
Sustainable Environment
    and Energy Systems
Middle East Technical University
    Northern Cyprus Campus
Kalkanli, Güzelyurt, Turkey

Seda Demirel Topel
Faculty of Engineering
Department of Material Science and
    Nanotechnology Engineering
Antalya Bilim University
Antalya, Turkey

Hadi Zahmatkesh
Department of Computer Engineering
Antalya Bilim University
Antalya, Turkey

Chapter 1

Introduction

Fadi Al-Turjman

Department of Computer Engineering, Antalya Bilim University, Antalya, Turkey

Content

1.1 Book Outline 2

Recent advances in mm-Wave/terahertz (THz) communications, as well as mobile access to computational power, are fostering a rapid growth of wearable and nanotechnologies. In particular, application of such technologies to healthcare can improve the control over health and well-being, providing motivation to achieve personal bests and creating a sense of community. By recording and reporting information about behaviors such as physical activity or sleep patterns, these technologies can educate and motivate individuals toward better habits and better health conditions. In addition, clinical and self-monitored data collected by wearable devices provides a means for improving the early-stage detection and management of diseases as well as reducing the overall costs of more invasive standard diagnostic approaches.

In this book, we aim to discuss some of the ongoing key innovations in communications and nanotechnologies involved in the emerging Internet of Nano-Things (IoNT) paradigm, which is setting the basis for the future of smart wearable devices and approaches. The manufacturing processes of some of these wearable technologies and their working principles are discussed in detail. In addition, we investigate several up-to-date communication protocols and design aspects that can dramatically affect performance.

Simply, this book aims to summarize recent achievements in the area of wearable and wireless body-area networks and point to existing challenges that the research and industrial communities are facing in trying to extend this paradigm to a broader spectrum of related devices and communication protocols. This will indeed provide solid guidelines for the future research directions and studies in this area.

Accordingly, our main contributions in this work can be summarized as follows:

- We start by a comprehensive overview about the millimeter waves (mm-Waves) and their modeling techniques. We further classify various candidate categories that can be applied in IoNT applications.
- We overview nanosensor technology and list the detailed challenges in manufacturing and data gathering.
- Various techniques and tools available for performance evaluation of the nanosensors used in wireless body area networks (WBAN) in IoNT environments are also presented in order to realize more energy-efficient solutions. These solutions and energy-related challenges are discussed in detail.
- We describe prominent performance metrics to understand how energy efficiency is evaluated. Then, we elucidate how energy can be modeled and harvested in the literature.
- We present energy-based routing protocols in critical applications such as healthcare. The potential reduction of the consumed energy and service capacity due to body mobility effects are considered, as well as other performance metrics such as delay, throughput, and resource utilization.
- We present the main motivations in carrying smart devices and the correlation between the user WBAN and the mobile application usage.
- We also propose a bio-inspired routing algorithm to construct, recover, and select reliable paths that tolerate the failure while satisfying quality of service (QoS) requirements.

1.1 Book Outline

The rest of this book is organized as follows. Chapter 2 focuses on the challenges introduced and strategies to be considered while dealing with the IoNT issues. In Chapter 3, we focus on the nanosensors and WBAN technology in IoNTs that are utilized in several industries. In Chapter 4, we focus on design aspects of the mm-Waves and their applications in the IoNT era. In Chapter 5, we put forward a data delivery framework for the nanoscale systems, where a number of nanosensors are disseminated over the human body and the like to help in disaster management. Chapter 6 proposes a resilient cache replacement approach for the WBAN based on the value of sensed information. In Chapter 7, energy efficiency in three common schemes, including the one-hop, cooperative, and two-hop networks are studied and a new method for a batteryless WBAN is proposed. In Chapter 8, we investigate a new routing technique for the IoNT paradigm in terms of energy consumption and hop counts. The existing routing protocols for the wireless nanosensor networks are reviewed in Chapter 9. A detailed classification and comparison of the existing protocols has been performed and a new simple energy-aware approach is proposed. To achieve the reliable data transmission, a low complexity parity check (LCPC) approach is proposed in Chapter 10 to increase the targeted WBAN efficiency by detecting and correcting common wireless signal errors. In Chapter 11, we conclude with an overview of the existing energy-scavenging methods and their potential usages in WBANs.

Chapter 2

Internet of Nano-Things (IoNT) & WBAN

Fadi Al-Turjman, Enver Ever, and Hadi Zahmatkesh
Department of Computer Engineering, Antalya Bilim University, Antalya, Turkey

Contents

2.1 Introduction 6
2.2 IoNT Market Opportunity in the 5G Era 7
2.3 IoNT Architecture in 5G 8
2.4 IoNT Design Factors and Assessment 9
    2.4.1 Short Wavelength 9
    2.4.2 Energy Harvesting 10
    2.4.3 Security 10
    2.4.4 Connectivity 11
    2.4.5 Delay 11
    2.4.6 Cost 11
2.5 IoNT Physical Layer and 5G 12
2.6 IoNT Communication Protocols and 5G 13
    2.6.1 Wireless Communication Models 13
    2.6.2 MAC Protocols 14
    2.6.3 Routing Algorithms 14
2.7 Constraint of the IoNT 15
2.8 Open Research Issues 17
2.9 Concluding Remarks 17
References 18

2.1 Introduction

The Internet of Things (IoT) has transformed the use of Internet in recent years with various applications. For instance, in the healthcare domain, body area networks (BANs) are used to collect crucial information of patients and send it to the computing systems of service providers in order to monitor a large number of people more efficiently and accurately. Moreover, especially in domains such as healthcare for elderly and physically disabled people, sensors deployed in the environment can provide useful mechanisms in terms of real-time monitoring and rehabilitation [1]. Cloud-based BANs are composed of (i) nodes for sensing, (ii) nodes that relay data in addition to their own sensing, and (iii) hubs or sink nodes that send control signals to gather information from the network and transmit to the cloud. Communication channels in BANs can be classified into two categories depending on the type of nodes: (i) on-body communication, which is communication of an on-body sensor node to a hub or another on-body node, and (ii) in-body communication, which is communication of an implant node to the hub or an on-body node. On-body communication in the latter can be categorized as line of sight (LoS) or non-line of sight (NLoS) depending on the relative position of the nodes. Physical and MAC layers' parameters, such as the packet and payload size, are specified by IEEE 802.15.6 standard and play a key role in the spectrum management and optimization [1, 2].

In addition to the advancements in Internet and sensing technologies, recent improvements in nanotechnology and design of nanoscale components (e.g. nanosensors, nano-antennas, etc.) have given rise to a new class of applications and services in various sectors and industries such as health [2] and agriculture [3], and have stimulated the evolution of a new networking paradigm called Internet of Nano-Things (IoNT). IoNT is defined as an interconnection of nanoscale devices with the current communication technologies and the Internet [4]. Terahertz (THz) band communication is utilized through new developments in areas such as spectrum management and antenna design to obtain data from various objects. All these developments in turn result in the discovery of novel applications. For instance, environmental nanosensors can provide information about allergens and pathogens in a given environment, while on-body nanosensors can collect electrocardiographic and other similar important signals [1]. By combining this information through IoNT, it would be much easier to monitor and diagnose a patient's conditions more accurately [1]. The IoNT paradigm is characterized by a very large number of nanodevices, technologies, and protocols. It is really important to take into account various properties for the IoNT, such as availability, flexibility, reliability, scalability, and interoperability, in the 5G era. These properties can easily be supported by the facilities of cloud computing [9]. Cloud computing can provide efficient solutions for communications in the nanoscale. It can facilitate quick setup and integration of nanodevices. Moreover, it can lower the deployment cost and data-processing complexity.

The aforementioned examples are just a few areas where IoNT and nanosensing technology have made great enhancements. On top of all the studies presented for efficient use of the new spectrum introduced for IoNT, it is necessary to investigate the challenges and opportunities introduced by the IoNT concept. In this chapter, we present an overview for the IoNT and focus on the strategies to be considered while dealing with the challenges introduced by the IoNT paradigm. Accordingly, our main contributions in this work can be summarized as follows:

- This chapter provides a critical overview of the IoNT considering the main application areas, architecture, limitations, and design factors.
- Potential enabling technologies from the physical layer up to various routing protocols that can be employed for the IoNT, as well as the interaction with the cloud-based infrastructures, are presented.
- Various challenges regarding terahertz spectrum management in 5G communication systems are comprehensively discussed.

The rest of this chapter is organized as follows. An overview of the IoNT market opportunity is provided in Section 2.2. Section 2.3 presents the general architecture of the IoNT. The most important factors in the design of IoNT are discussed in Section 2.4. IoNT physical layer communication and communication protocols, including various MAC approaches and routing algorithms, are presented in Sections 2.5 and 2.6, respectively. Section 2.7 presents some of the constraints caused by nanocommunications in the IoNT paradigm. Section 2.8 gives future research directions and discusses some open research issues. Finally, Section 2.9 concludes this chapter.

2.2 IoNT Market Opportunity in the 5G Era

The number of connected devices is expected to rapidly increase in the coming years. Further growth is also expected in the network size and complexity for real-time traffic monitoring, since most of the communication devices are smart devices with multiple features. In order to enable interaction between these devices, communication is required between real-world physical elements. The IoNT will facilitate communication over the Internet for such devices.

According to market research analysis, it is estimated that the global IoNT will grow at a significant compound annual growth rate (CAGR) of over 24% from 2016 to 2020 [7]. The increasing utilization of nanotechnology is one of the most important factors leading to the growth of IoNT markets on a global scale. Significant investments in research and development by private as well as public sectors have great potential to lead to notable growth for the IoNT market in the coming years. For example, the NMP program under the Seventh EU Framework Programme invested approximately USD 480 million in nanomedicine research projects such as regenerative medicine, nanodiagnostics, and targeted nanophar-maceuticals in 2014 [7]. Through commercialization of nanomaterials, such as nanoscale electronic memory, the IoNT market is expecting dramatic growth.

The healthcare segment, among many other segments (e.g. manufacturing, transportation and logistics, and energy and utilities) in the IoNT market, was the largest as of 2015, with about 40% market share [7]. This segment is estimated to have a significant growth from 2016 to 2020 due to the government plans for digitalization of healthcare operations.

Some of the crucial factors leading to the market growth mentioned are tech-nological innovations and the entry of new manufacturers. Manufacturers have to pay attention to interoperability issues and develop test-beds, which are good representations of real scenarios, in order to be successful in this market. Some of these key vendors in the IoNT market are Alcatel-Lucent, Cisco, IBM, Intel, and Qualcomm.

2.3 IoNT Architecture in 5G

Understanding the architecture of IoNT helps us to obtain a clearer insight of the functionalities involved. The IoNT should be capable of interconnecting billions of nanosensors and nanodevices through the Internet, interacting with each other in a distributed manner. Regardless of the application of the IoNT, the general network architecture of the paradigm contains the components [1, 4] presented in Figure 2.1.

Nanonodes: These are the end-points such as nanosensors and nano-actuators, which are able to perform simple computation and processing tasks. Due to their limited communication capabilities, reduced energy, and limited memory, they can only transmit over very short distances. An example of nanonodes is nanomachines having communication capabilities combined in various types of things, such as books and keys. These devices can also be used inside the human body as biological nanosensors.

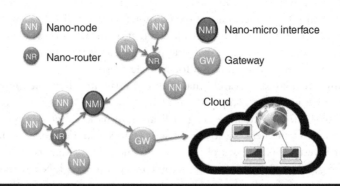

Figure 2.1   **Network architecture and main components in the IoNT.**

Nanorouters: Compared to the nanonodes, nanorouters have larger computational resources, and they are appropriate for collecting information from limited nanomachines (e.g. nanosensors). Moreover, nanorouters can also monitor the behavior of nanonodes by sending very simple control commands such as sleep, on/off, etc. However, this would increase their size, and consequently their deployment would be more invasive.

Nano-micro interface devices: Information forwarded by nanorouters is aggregated by nano-micro interface devices. These devices can handle information from microscale to nanoscale devices. Nano-micro interfaces can be considered as hybrid devices that are able to communicate in nanoscale using nanocommunication technologies. In addition, they can also use classical communication models in micro/macro communication networks.

Gateways: These are used to enable the remote monitoring of the entire system over the Internet. Gateways can receive the information from a nano-micro interface and forward it to the related service provider. For example, in the case of a healthcare monitoring system, all the sensor data from the human body can be forwarded over the Internet to the healthcare provider.

The IoNT architecture can also be customized according to the applications and objectives of the network in order to achieve the specific goals of a system. For example, in [5], the nanorouters forward the collected data to cognitive relay nodes that are usually connected to the Internet for remote processing. These cognitive nodes act and make decision based on the nanonetwork conditions in order to save considerable amounts of energy in the entire system.

2.4 IoNT Design Factors and Assessment

The fifth generation of wireless networks (5G) is expected to be available very soon. One of the main objectives specified is to have ubiquitous communication anytime and anywhere between anyone and anything. In this section, the most important design factors of the IoNT paradigm that will significantly affect the energy and spectral efficiency, and quality of service (QoS) are discussed [22].

2.4.1 Short Wavelength

In existing implementations, antennas and graphene transceivers are employed popularly [15]. However, although this provides potentially good data rates with frequency in the range of 0.1 to 10 terahertz, because of the very low wavelength, the practical range is reduced to around 10 mm [16]. Therefore, the limitations and opportunities of using very high frequency signals should be considered in the design of the IoNT. For example, in the case of healthcare monitoring systems, one of the promising solutions for the data exchange is electromagnetic-based communication at the terahertz band [19]. This would significantly contribute to potential medical technologies in the utilized spectrum at the terahertz band because it is less susceptible to propagation effects (e.g. scattering) and it also has safety advantages for biomedical tissues [19]. Applications similar to the ones presented in [19] can effectively make use of terahertz band communication, since the limitations in terms of the range and the line of sight do not significantly affect the nature of the application. However, if an application requires significant enhancements in terms of coverage, for example, in case mobility is required for the application, the allocated spectrum becomes a serious limitation.

2.4.2 Energy Harvesting

Energy harvesting is a crucial factor in the IoNT paradigm in the 5G era. For example, in the case of nanoscale batteries, they cannot store much energy for the long duration of nonstop operation of the nanonodes. Due to the fact that nanonetworks' sensors usually come with limited capabilities and battery power, the implementation and data routing processes face serious challenges. Thus, data delivery algorithms and protocols should be energy efficient to extend the overall network lifetime, which is directly proportional to the nanonode's battery level.

Recent studies and surveys have emphasized the demand for microscale energy-scavenging technology due to its importance in functioning ultra-low-power electronic circuits. However, it is essential to identify the sources of energy scavenging technology to evaluate whether the corresponding scavenging technology is useful for a particular application in the IoNT paradigm or not. In general, solar, thermal, mechanical movement or vibration, and ambient radio-frequency (RF) sources can be used to extract energy from microscale energy scavenging circuits. They can be divided into three major categories: electrostatic, electromagnetic, and piezoelectric [21]. According to this classification, the energy extraction capacity, physical size, robustness, and output impedance characteristics of the harvester circuits can vary.

2.4.3 Security

One of the most important challenges as a result of the growth of the IoNT market is related to the security of data communicated over the Internet. For example, in the healthcare domain, a bio-cyber-attack can steal people's personal health-related information. This information can be used to create new types of viruses to hack into already-deployed nanosensors in the IoNT. Therefore, security assurance methods should be applied to communication networks in the 5G era, especially in the IoNT, in order to prevent such problems, considering the nature of IoNT communications carefully.

2.4.4 Connectivity

The IoNT defines a paradigm where all types of nanodevices (e.g. nanosensors, nanoactuators, etc.) are connected to the nanonetworks and are able to interact with each other in real time. This form of seamless connectivity is essential for enabling the applications that involve IoNT technologies. In turn, connectivity becomes an important factor for integration of IoNT in the 5G environments as well.

2.4.5 Delay

In nanonetworks, the gateway must integrate data from different nanodevices (e.g. nanosensors). However, due to the timing difference in data propagation between nanosensors, this can lead to long delays for data before reaching the sink node. Therefore, a time-delayed data fusion method should be applied at the gateway for the processing of data before they are transmitted over the Internet.

2.4.6 Cost

Cost is another crucial factor in the design of IoNT in the 5G environment, and it has to be carefully considered by both users and service providers. Using cloud technology in the IoNT era can lower the cost for deployment and complex data processing while facilitating quick setup and integration of new nanodevices [9].

In the following, we discuss a number of studies that consider the aforementioned factors while designing the network. In [10], the authors propose a routing algorithm that can be dynamically deployed within a nanonetwork. They also examine static and dense topologies with several identical nodes. With the use of simulation, they considered energy harvesting and cost as the preliminary metrics for the assessment of the proposed algorithm.

The authors in [11] propose a routing algorithm for multi-hop data transmission, which is enabled by the latest developments in physical network layer coding. Energy harvesting and cost are considered as the primary design factors of the study in [11]. A channel-aware forwarding scheme for electromagnetic-based wireless nanosensor networks is proposed in [12], providing a solution to the terahertz frequency selective feature from the networking point of view. Energy harvesting and security are assessed as the preliminary design metrics in the proposed algorithm in [12]. In [15], an in-depth overview of nanosensor technology and electromagnetic communication among nanosensors is provided by considering energy harvesting, security, and terahertz channel modeling. Security related issues in nanocommunication, especially for molecular communication, are thoroughly discussed in [17] as well. The authors in [16] propose an on-demand probabilistic polling scheme in order to avoid unnecessary resource consumption in electromagnetic-based wireless nanosensor networks. They achieve higher bandwidth efficiency of IoT backhaul compared to the previously proposed algorithms.

Table 2.1 Summary of Design Factors

Ref	Sw	Eh	S	Cn	D	C
[10]	X	X	-	-	-	X
[11]	X	X	-	-	-	X
[12]	X	X	-	-	X	-
[15]	X	X	X	-	-	-
[16]	X	X	-	-	X	-
[17]	-	X	X	-	-	-

Sw: Short wavelength, Eh: Energy harvesting, S: Security,
Cn: Connectivity, D: Delay, C: Cost

- = Not Considered, X = Considered

A summary of the aforementioned design factors is presented in Table 2.1. According to this table, there is a significant need for a new IoNT-based solution that takes into consideration all of these factors to cope with the emerging cloud challenges and 5G spectrum optimization. Furthermore, it is most important to pay more attention to connectivity factors while optimizing the spectrum utilization in 5G-based solutions as highlighted in Table 2.1.

2.5 IoNT Physical Layer and 5G

The communication in nanonetworks can utilize nanomechanical, acoustic, electromagnetic, and chemical or molecular communication. Comparison for the existing communication technologies can be seen in Table 2.2. The physical signaling is at the THz levels; therefore, due to the necessary antenna sizes, special modulation techniques are required. On the other hand, research studies on use of graphene-based plasmonic materials for antennas to overcome signaling difficulties look promising. For example, in [6], the authors propose a graphene-based

Table 2.2 Comparison of Communication Technologies

Communication Type	Internet	Nano-molecular	Nano-wireless
Signal type	EM	Chemical	EM
Signal speed	High	Low	High
Power consumption	High	Low	Low

EM = Electromagnetic

plasmonic nano-antenna for communication between nanodevices. They reveal that by utilizing the high wave compression mode of surface plasmon polariton (SPP) waves in armchair graphene nanoribbons (AGNRs), graphene-based nano-antennas can operate at much lower frequencies than traditional metallic antennas of the same size.

The large bandwidth in terahertz band communication enables very high-speed communication which is envisaged in 5G wireless communication systems [8]. Moreover, the terahertz band offers a great amount of spectrum resources, which in turn reveal the potential to support data rates of up to even 1 Tbps [4]. In addition, Multiple input multiple output (MIMO) techniques can be incorporated in the terahertz-band communication in order to increase the data throughput and improve the reliability of the systems [8]. The frequency spectrum of terahertz band is already being investigated in studies such as [20] for 5G communication systems.

Existing studies investigate ways of efficient spectrum allocation through pushing the carrier frequencies into the terahertz band quite extensively. Various antenna designs are proposed for small cells and small coverage areas for this frequency spectrum [20]. QoS related challenges are different when compared to traditional microwave spectrums using larger-range cellular infrastructures. To start with, the interference structure in terahertz spectrums using systems can be principally different from what is so far observed at lower frequencies. This structure causes various limitations, which include the need for LoS links, since reflections will deflect the waves and molecular absorption would significantly affect the signal strength. Therefore, the following challenges are quite important for terahertz spectrum management in 5G communication systems:

- Implementations of antennas with high directivity to transmit and receive
- Solutions for molecular absorption caused by short wavelength
- Solution for blockage of high-frequency radiation

2.6 IoNT Communication Protocols and 5G

2.6.1 Wireless Communication Models

Communication in nanoscale can be classified in molecular communication and nano-electromagnetic communication. Molecular communication is defined as the transmission and reception of information encoded in molecules, while transmission and reception of electromagnetic (EM) radiation from components based on nanomaterials defines nano-electromagnetic communication. It is of great importance to study the communication nature in very short range, since it is functioning at the nanoscale. Therefore, the assumed path loss formula shall be at the terahertz level. This formula usually has two essential parts: the absorption and the spread path loss.

In [13], four different power spectral densities (p.s.d.) were studied, namely, optimal p.s.d., flat p.s.d., the Gaussian pulse, and the p.s.d. for the case of the transmission window at 350 GHz. The authors concluded that for the very short communication range in bio-inspired applications, transmission rates can be supported up to terabits per second, indicating the promising future of the nanocommunication.

There are three types of data traffic that a nanonetwork can experience in an assumed IoNT paradigm: on-demand, periodic, and emergency traffic. Each of these are associated with a different quality of information (QoI) requirement based on the served IoNT application. The QoI requirement for the routed data can be decided by the following attributes: (1) the network reliability (fairness), (2) the end-to-end delay/latency, and (3) the energy consumed. Some of the recent studies such as [18] focus on MAC/scheduling algorithms, not only to provide high network throughput for the IoNT, but also to address challenges imposed by the limited memory of nanodevices, as well as energy consumption and harvesting over terahertz channels.

2.6.2 MAC Protocols

In the IoNT paradigm, the routing protocol should be coupled to the MAC layer through a cross-layer design. Feedback from the MAC and physical layers, in addition to the residual energy and current load of the IoNT nodes, will be utilized to identify and bypass critical links. Therefore, the network lifetime can be effectively prolonged by preserving residual energy and increasing network throughput, which can be achieved through load balancing. Reducing the ratio of lost packets during channel impairments is an important reliability objective, as well. Since classical MAC protocols typically limit the number of backoffs and retransmissions, they cannot directly be used in the IoNT [18]. This is because they do not capture the peculiarities of the terahertz band [13], such as the very large distant-dependent bandwidth and the very high path loss. Moreover, traditional MAC protocols do not capture the limited processing capabilities and the small-capacity memory as well as the temporal energy fluctuations of nanodevices [18]. Therefore, there is a need to design new MAC protocols and propose new solutions related to the IoNT paradigm.

2.6.3 Routing Algorithms

Numerous IoNT design aspects that stem from its unique features need to be incorporated into its routing protocols in order to realize the IoNT paradigm. Different challenges against routing protocols' design in terms of energy are still being investigated. Traditional nanonetworks spend energy in almost all processes. They spend energy while making data transmissions and data sensing as well as data processing. There have been a few attempts toward achieving energy efficiency in such networks via wireless multi-hop networking such as [13] and [14]. However, such schemes are only applicable in static wireless networks and are impractical in multi-hop nanoscale wireless networks with random topologies. Routing algorithms for the IoNT can be classified to cluster-based algorithms vs. cognitive-based algorithms. For example, a routing scheme for energy harvesting in terahertz bands was proposed in [14]. The routing scheme assumes a hierarchical cluster-based architecture. Packet transmission from the source to the cluster-head or nanocontroller can be direct or multi-hop based on the probability of saving energy through transmission, optimizing throughput, and minimizing nanosensors' load. On the other hand, in [5], a cognitive-based routing algorithm was proposed for data delivery in nanoscale networks where nanorouters forward the collected data to cognitive relay nodes for acting and making decisions based on the network conditions.

2.7 Constraint of the IoNT

It is possible to classify the constraints of the IoNT into primary vs. secondary constraints. Primary constraints are the most important limitations of the network in order to achieve accurate results. Secondary constraints have less importance compared to the primary ones, but still can significantly affect the performance of the network under study. Energy consumption is one of the most important issues in implementing the concept of IoNT. In the IoNT paradigm, energy is enough for transmitting one packet per each energy charge. Hence, retransmission is very expensive. Moreover, nodes require roughly 10 seconds to recharge. Some methods compose clusters and assign IP addresses for the cluster head while others do not assume IP addressing schemes. Most proposals can be divided into two types: (1) flooding-based routing schemes, and (2) peer-to-peer or point-to-point methods.

Another important constraint regarding the implementation of the IoNT is related to the deployment issue. For instance, in [10], the authors assume a grid-static deployment. They map this deployment into circular/radial deployment with the following assumptions:

- The routing protocol is built around a beacon node which is chosen externally; nodes on the same circle are considered same hop distance away from the beacon.
- Nodes on two consecutive circles are on the same radial.
- Nodes may be on radial, center of circles (beacon), on a circle, or between circles.

This scheme is engaged at two phases: the deployment phase and the data routing phase. In the deployment phase, a beacon sends two types of setup messages to all nodes: a low-power transmission node, with transmission coverage assumed to cover at least two circles (one radial distance), and normal-power transmission mode assuming to cover neighbors on one circle. For the deployment of low-transmission power, a setup message is sent by the beacon with the number-of-hops field set to 0 to all nodes. Any node that receives the message increases the number-of-hops field, and checks its energy to decide to participate in the topology as a relay or just a user node. A similar procedure is carried out for the normal power transmission. Finally, a node can be in either a normal-power transmission mode or a low-power transmission mode. In the data-routing phase, any node receives a packet; then it checks its n-hop count against the packet-hop count and its mode. If it is low-power mode and its n-hop count is within the source-hop-count and the recover-hop-count, and it is a relay node, then it forwards the packet. If not, it checks the following. If it is in normal power mode, and if its hop count is equal to either the source-hop-count or the receiver-hop-count, then it can forward the packet. Otherwise, it refrains.

In [11], a greedy routing algorithm is proposed by defining three circles with diameters r_d, r_min, and r_max, respectively. The algorithm routes the packet in a pair-based routing as follows. The source node composes a pair with the closest neighbor within r_d distance, which we call a source-neighbor pair. The algorithm then checks the distance of the source-neighbor pair with the destination. If it is less than r_max, it finds the closest neighbor v to the destination and pairs it with the destination. Let's call this pair, the *pair_next*. The algorithm then forwards the packet on a pair to pair bases. Otherwise, it finds *pair_next*, such that r_min is less than or equal to the distance between the *pair* and its associated *pair_next* that is also less than or equal to r_max. Then it forwards the packet. This process is repeated until the packet is delivered.

A centralized algorithm is proposed in [12] where the sink assigns a selection index value for each neighboring node to the source. The selection index depends on the distance between the source and the relay node and the attainable channel capacity. The sink chooses the relay node with the highest selection index to forward the message. The scheme assumes that the sink node knows the location of each nanonode and the channel capacity of each link.

When secondary constraints are considered, radiation exposure evolving from the communication/routing techniques must be considered carefully, especially in the biomedical applications of the IoNT paradigm. Electromagnetic fields (EMFs) at different spectrums can negatively affect the human body in different ways. For example, the spectrum between 1 MHz to 10 GHz frequency can penetrate through tissues and produce absorbable heat. However, EMFs over 10 GHz are strongly blocked by the skin, and the heat created causes damage such as eye cataracts or skin burns if the field density is above 1000W/m2. Accordingly, the International Commission on Non-Ionizing Radiation Protection (ICNIRP) developed general guidelines for EMF exposure limits recognized under the World Health Organization (WHO). And thus, the shared spectrum RF emissions must be considered carefully to avoid undesired human body damage. Security and privacy concerns and the high cost of initial investment are also some of the secondary constraints in implementing the IoNT.

2.8 Open Research Issues

In spite of various research activities and rapid developments that have been achieved in recent years regarding the IoNT paradigm, there are still many challenges and open research issues that need to be taken into account regarding the performance improvements of the IoNT. One of them is terahertz band channel modeling. The IoNT needs to transmit very large amounts of data in a timely and reliable manner. Therefore, the impact of molecular absorption on the path loss and noise should be accurately analyzed. This will help to locate the best transmission window in terms of achievable information rates and channel capacity. Moreover, the impact of multi-path propagation on the capacity and achievable information rates should be accurately investigated.

Another important challenge is related to the MAC protocol. The terahertz band supports very high bit rates and has a specific relation between the available transmission window, the bandwidth for each window, and the transmission distance. Therefore, research into transmission schemes would be beneficial in order to develop novel transmission techniques using the relation between the transmission bandwidth and the transmission distance. The MAC protocol should also guarantee that the transmitter and receiver are properly aligned before the transmission of the data packet.

Another key challenge in the IoNT paradigm is security. Research for mechanisms to develop new authentication schemes for nano-things as well as to guarantee data integrity and user privacy is essential in order to protect important personal information of the people. It is also important to investigate neighbor discovery mechanisms that utilize the high directivity of the terahertz antennas to specify the relative location and orientation among various nano-objects.

2.9 Concluding Remarks

Nowadays, the emerging paradigm of the IoNT is rapidly growing and aiming to improve people's quality of life. These enhancements in the IoNT and nano-technology will have a great impact on advanced development in various fields, such as healthcare, agriculture, environmental monitoring, and next-generation cellular systems, to just name a few. In this chapter, we provided an overview of the IoNT, considering the architecture and main application areas. We also presented potential enabling technologies from the physical layer up to various routing protocols that can be employed for the IoNT. We discussed challenges regarding terahertz spectrum management in 5G communication systems. Moreover, radiation exposure and other side effects for the IoNT applications were discussed.

References

1. Balasubramaniam S., Kangasharju J. 2013. Realizing the internet of nano things: Challenges, solutions, and applications, *Computer*, 46(2), 62–68.
2. Hasan M.Z., Al-Turjman F. 2018. Analysis of cross-layer design of quality-of-service forward geographic wireless sensor network routing strategies in Green Internet of Things, *IEEE Access Journal*, 6(1), 20371–20389.
3. Bhargava K., Ivanov S., Donnelly W. September 2015. Internet of nano things for dairy farming. In *Proceedings of the Second Annual International Conference on Nanoscale Computing and Communication*, 24.
4. Akyildiz I.F., Nie S., Lin S.C., Chandrasekaran M. 2016. 5G roadmap: 10 key enabling technologies. *Computer Networks*, 106, 17–48.
5. Al-Turjman F. 2017. A cognitive routing protocol for bio-inspired networking in the Internet of nano-things (IoNT). *Mobile Networks and Applications*, 1–15.
6. Jornet J.M., Akyildiz I.F. 2013. Graphene-based plasmonic nano-antenna for tera-hertz band communication in nanonetworks. *IEEE Journal on Selected Areas in Communications*, 31(12), 685–694.
7. Global Internet of Nano Things Market 2016–2020. 2016. https://www.technavio.com/report/global-it-professional-services-internet-nano-things-market. Accessed on 14 May 2018.
8. Khalid N., Akan O.B. 2016. Experimental throughput analysis of low-THz MIMO communication channel in 5G wireless networks. *IEEE Wireless Communications Letters*, 5(6), 616–619.
9. Al-Turjman F., Alturjman S. 2018. Context-sensitive access in Industrial Internet of Things (IIoT) healthcare applications. *IEEE Transactions on Industrial Informatics*, 14(6), 2736–2744.
10. Liaskos C., Tsioliaridou A., Ioannidis S., Kantartzis N., Pitsillides A. May 2016. A deployable routing system for nanonetworks. *IEEE International Conference on Communications (ICC)*, 1–6.
11. Zhou R., Li Z., Wu C., Williamson C. July 2012. Buddy routing: A routing para-digm for nanonets based on physical layer network coding. *21st IEEE International Conference on Computer Communications and Networks (ICCCN)*, 1–7.
12. Yu H., Ng B., Seah W.K. September 2015. Forwarding schemes for EM-based wire-less nanosensor networks in the terahertz band. *Proceedings of the Second Annual International Conference on Nanoscale Computing and Communication*, 17.
13. Jornet J.M., Akyildiz I.F. 2011. Channel modeling and capacity analysis for electro-magnetic wireless nanonetworks in the terahertz band. *IEEE Transactions on Wireless Communications*, 10(10), 3211–3221.
14. Pierobon M., Jornet J.M., Akkari N., Almasri S., Akyildiz I.F. 2014. A routing framework for energy harvesting wireless nanosensor networks in the terahertz band. *Wireless Networks*, 20(5), 1169–1183.
15. Akyildiz I.F., Jornet J.M. 2010. Electromagnetic wireless nanosensor networks. *Nano Communication Networks*, 1(1), 3–19.
16. Yu H., Ng B., Seah W.K. 2017. On-demand probabilistic polling for nanonetworks under dynamic IoT backhaul network conditions. *IEEE Internet of Things Journal*, 4(6), 2217–2227.
17. Dressler F., Kargl F. 2012. Towards security in nano-communication: Challenges and opportunities. *Nano Communication Networks*, 3(3), 151–160.

18. Akkari N., Wang P., Jornet J.M., Fadel E., Elrefaei L., Malik M.G.A., Almasri S., Akyildiz I.F. 2016. Distributed timely throughput optimal scheduling for the Internet of Nano-Things. *IEEE Internet of Things Journal*, 3(6), 1202–1212.
19. Abbasi Q.H., Yang K., Chopra N., Jornet J.M., Abuali N.A., Qaraqe K.A., Alomainy A. 2016. Nano-communication for biomedical applications: A review on the state-of-the-art from physical layers to novel networking concepts. *IEEE Access*, 4, 3920–3935.
20. Barros M.T., Mullins R., Balasubramaniam S. 2017. Integrated terahertz communication with reflectors for 5G small-cell networks. *IEEE Transactions on Vehicular Technology*, 66(7), 5647–5657.
21. Al-Turjman F. 2017. 5G-enabled devices and smart-spaces in social-IoT: An overview. *Elsevier Future Generation Computer Systems*, DOI: 10.1016/j.future.2017.11.035.
22. Al-Turjman F. 2018. QoS–aware data delivery framework for safety-inspired multimedia in integrated vehicular-IoT. *Elsevier Computer Communications Journal*, 121, 33–43, 2018.

Chapter 3

Nanosensors for the Internet of Nano-Things (IoNT): An Overview

Seda Demirel Topel[1] and Fadi Al-Turjman[2]

[1]*Faculty of Engineering, Department of Materials Science & Nanotechnology Engineering, Antalya Bilim University, Antalya, Turkey*
[2]*Department of Computer Engineering, Antalya Bilim University, Antalya, Turkey*

Contents

3.1 Introduction 22
3.2 Understanding the Sensing Mechanism at Nanoscale 22
3.3 Examples of Research Studies Targeting the Internet of Nano-Things
(IoNT) in Healthcare Applications 23
    3.3.1 IoNTs for the Patient's Surrounding Environments 24
    3.3.2 IoNTs for the Patient's Body 27
3.4 Nanosensor Fabrication and Material Science 29
    3.4.1 Optical Nanosensors 30
    3.4.2 Electrochemical Nanosensors 33
    3.4.3 Mechanical Nanosensors 35
3.5 Nanosensor and Energy Harvesting 35
3.6 Nanosensors and Body Networks 36
3.7 Open Research Issues 38
3.8 Concluding Remarks 39
References 40

3.1 Introduction

Fifty years ago, the famous physicist Richard Feynman gave a talk called "There Is Plenty of Room at the Bottom" at an annual meeting of the American Physical Society at Caltech. With his historic speech, he laid the foundations of a new era called Nanotechnology and emphasized the importance of nanotechnology with these words: "I don't know how to do this on a small scale in a practical way, but I do know that computing machines are very large; they fill rooms. Why can't we make them very small, make them of little wires, little elements, and by little, I mean little?" By now, the scientists have made the dream possible within the last three decades.

Nowadays, nanotechnology has become one of the most significant research topics, and it promises novel solutions for several applications in the industrial, military, healthcare, biomedical, environmental, agricultural and textile sectors. Each of those application fields requires new nanodevices which acquire, generate, compute, process, and transfer the data at a nanoscale dimension. These nanodevices are interconnected with the existing communication systems, which produces a new domain that is called the Internet of Nano-Things (IoNT). The nanodevices used in IoNTs are mainly composed of nanosensors in order to supply the communication between the system and the device. In IoNTs, nanosensors are connected to the Internet via local gateways. Integrating IoNT with other local area networks of these nanosensors can significantly expand the range of services which can be provided to public users as well as decision makers.

This chapter has been mainly focused on the nanosensors in IoNTs that are utilized in several industries, and it has been organized as follows: In Section 3.2, we aim at understanding the sensing mechanism at the nanoscale level. In Section 3.3, examples of the existing research studies targeting the Internet of Nano-Things in healthcare and medicine applications are proposed. The nanosensor fabrication process and its material science are discussed in Section 3.4. In Section 3.5, nanosensor and energy-harvesting issues are investigated. In Section 3.6, the nanosensor and its relation to body area networks is detailed. Open research issues and concluding remarks are highlighted in Sections 3.7 and 3.8, respectively.

3.2 Understanding the Sensing Mechanism at Nanoscale

A sensor is a device that detects and responds to physical, chemical, or biological stimuli. When the nanotechnology merges with the sensor technology in order to sense with a nanomaterial having the size of 10 to 100 nanometers, it is called a "nanosensor." The nanosensor is much more efficient than a regular sensor due to the reduced size of the sensor material. So what happens when the dimensions of a material are in the nanometric scale? The modifications in the properties due to reduction in grain size to nanoscale dimensions are very large, and in most cases the resultant properties are superior to those of conventional materials. It can be well

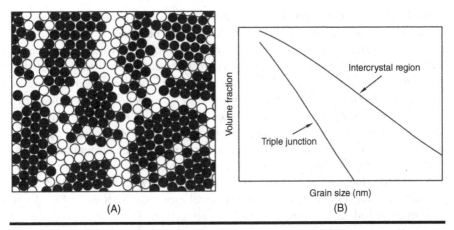

Figure 3.1 **(A) The hypothetical structure of a nanomaterial. The black circles indicate atoms in the grain, while the white circles indicate atoms at the grain boundaries. (B) Increase in the intercrystalline region (grain boundaries) and triple junctions with decrease in grain size of nanomaterials [1].**

explained by the grains in the nanostructured materials. Nanostructured materials are composed of grains and grain boundaries. Nanometer-sized grains contain only a few thousand atoms within each grain. A large number of atoms reside at the grain boundaries (Figure 3.1A). As the grain size decreases, there is a significant increase in the volume fraction of grain boundaries or interfaces and triple junctions (Figure 3.1B). When the fraction of atoms residing at defect cores, such as dislocations, grain boundaries, and triple junctions, becomes comparable with that residing in the core, the properties of the material are restricted by the defect configurations, dynamics, and interactions. Hence, the mechanical and chemical properties of nanomaterials are significantly altered due to defect dynamics [1].

The theory mentioned above is one of the fundamental explanations of how performance is related to sensitivity and specificity of the nanosensor. The other feature that can enhance the performance of the nanosensor is the ability to function with ligands offering a single molecule selectivity [2], and this offers more rapid and multiplexed detection of analytes in integrated nanosensors systems [3–5].

3.3 Examples of Research Studies Targeting the Internet of Nano-Things (IoNT) in Healthcare Applications

As we aforementioned, nanotechnology and IoNTs have significant roles in myriad applications. In particular, the healthcare and medicine industries hold the potential to bring the greatest benefits to society. As Metin Sitti said, small-scale networks have quite a bright future, especially in healthcare and bioengineering scenarios, because the corresponding devices in the network are "unrivalled for accessing into small, highly confined and delicate body sites, where conventional medical devices fall short without an invasive intervention" [6]. In light of this explanation, the combination of medicine with IoNTs will result in revolutionary changes in disease prevention, diagnosis, and treatment. Moreover, the early-stage diagnosis of the diseases and targeted therapy will save a great number of lives as well as extended the average lifespan. Therefore, IoNTs have opened a whole new world of potential remarkable betterments in medicine and healthcare. Additionally, it is forecasted that the global nanotechnology market will grow at a compound annual growth rate (CAGR) of around 17% during the 2018–2024 period, while IoNTs and medicine markets will grow at CAGR of 22.81% from 2016 to 2020 and 12.3% from 2013 to 2019, respectively [7].

Current healthcare applications are carried out in two domains: (1) the patient's surrounding environment, and (2) the patient's body area. In the first domain, sensors and actuators can be installed in the patient's surrounding environment in order to track the patient's daily activities. However, in the second domain, the nanosensors can be planted inside the body via medical operations and/or nano-drugs given orally to the patient or through the blood vessel. Then, they monitor the biological properties of the tissues and organs, and send the collected data to a gateway. And thus, they can detect significant diseases such as cancer and other health issues caused by viruses, chemicals, and/or heart attacks.

3.3.1 IoNTs for the Patient's Surrounding Environments

Due to aging and chronic diseases, public health expenditures have increased in many countries. Therefore, shifting from expensive institutional care centers to home, personal well-being, and preventive healthcare is highly promoted. In order to realize this shift, there is a need for long-term monitoring of the physiological signals. New nanosensor technology offers solutions for long-term monitoring with its small, reliable, comfortable, flexible, personalized, power-efficient, and safe properties.

Current medical sensors and remote-control monitoring units are generally based on electrodes embedded into electronic devices connected with wires. In addition, with the developments in microelectronic and communication technologies, the embedded smart-device networks have been produced in order to sense their environment, process data, and exchange information by forming networks with the concept of IoNTs [8, 9]. This technology, including IoNTs, has been integrated with other enabling technologies such as the radio frequency identification (RFID), near-field communication (NFC), global positioning systems (GPS), utra-wide band (UWB) communication, sensors, and actuators [9, 10]. These technologies provide a unique and secure identification for the physical item, including the measurement, and management processes for rapidly changing data in

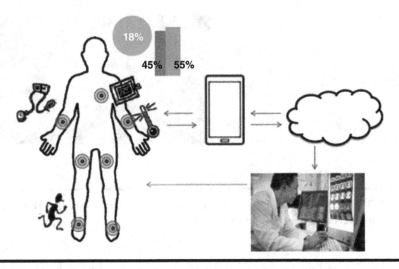

Figure 3.2   An application of remote healthcare using IoNTs [11].

real-time fashion. And thus, IoNTs have been applied in home environment for residential healthcare applications, which makes it possible to connect everything at any time and any place. In this regard, the nanosensor measures the physiological parameters and activities such as blood pressure, muscle activity, heart activity, glucose level, and body temperature, etc. Then the recorded parameters are sent to a mobile device or to the cloud for storage and extensive analysis. Afterward, physicians can better diagnose the patient case and provide a long-term feedback as depicted in Figure 3.2.

In 2014, a research group from Finland developed an integrated system for wireless physiological monitoring [11]. This integrated system was built from the combination of an an inkjet printer and a radio frequency (RF) system on chip technologies. In this study, Sillanpää and coworkers manufactured the biosensor on a single substrate. The wiring and the antenna were inkjet-printed with silver (Ag) nanoparticle ink, and the core sensor was a commercial NRF51822 system on chip (SoC). The resultant device has been used to measure electrocardiogram (ECG) signals and send the signals to a smartphone (Figure 3.3). Furthermore, the mobile device could use the Internet connection to send the raw data (heart rate, breathing rate, etc.) to a cloud service and/or doctors. The device can read up to a distance of 14 meters using the printed antenna in a close vicinity of the body [11].

Besides the silver nanoparticles, there are other types of nanomaterials such as polypyrrole, graphene, copper, gold nanoparticles, and carbon nanotubes (CNTs) which can be easily printed on flexible plastic, textile, paper, glass, and metallic surfaces in order to fabricate inkjet-printed RFID tag sensors for wearable applications for healthcare [12–17]. The fabrication of electronic devices by

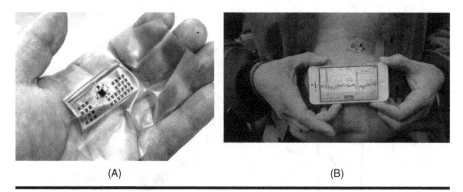

(A)   (B)

Figure 3.3   (A): The sensor fabricated on a stretchable substrate. (B): The sensor signal is delivered in real time in mobile device [11].

inkjet-printing technique is inexpensive compared with existing chemical vapor deposition (CVD), plastic vapor deposition (PVD), atomic layer deposition (ALD), and lithography. The flexible plastic substrates of polyethylene terephthalate (PET), polydimethylsiloxane (PDMS), polymethyl methacrylate (PMMA), and polyimide (PI) have been widely used for developing flexible and strechable electronic devices. The well-defined, high-resolution, and conductive patterns can easily be fabricated with inkjet printing on different flexible and strechable substrates for wearable electronics.

In a recent study, Kassal and co-workers demonstrated a low-power RFID tag sensor for potentiometric sensitivity [18]. The memory chip in the RFID tag has the capability to measure and eventually store the electrode potential, which is later on wirelessly transferred to a smartphone by near-field communication (NFC). Figure 3.4 shows the RFID/NFC-based chemical tag sensor platform and its operating principle. The RFID/NFC chemical tag sensor is suitable for detecting pH or ion-selective electrodes as part of a chemical sensor network for IoNTs. The practical application of RFID/NFC tag sensors was verified for milk spoilage by monitoring the pH value of souring milk over a period of six days. The pH of the souring milk decreased to 2.4 for the first two days due to lactic acid formation. Thereafter, the pH value stabilized at 4.3, which was in agreement with the 4.28 pH value measured by a laboratory meter. Furthermore, a buffer solution of pH 6.00 was monitored for five days where RFID tags recorded data at an interval of every ten minutes and transferred the data to the PC. The measurements showed the fluctuation of pH value between 5.89 and 6.10 over the five days, averaged to a pH of 6.03. Therefore, RFID/NFC tag sensors show potential for IoNT applications.

Another example of nanosensors for healthcare applications is the continuous glucose monitoring systems. These systems have been attached to the patient's body

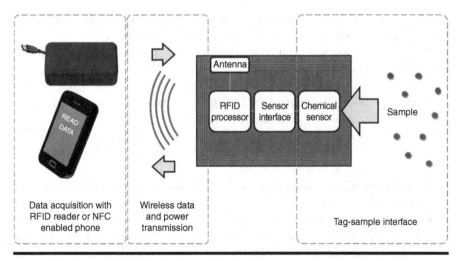

Figure 3.4 Schematic illustration of an RFID-based tag chemical sensor platform and its operating principle [18].

in the form of plaster and regularly (between one to five minutes), they measure the glucose level over a time span of five to fourteen days [43]. The system is composed of a glucose sensor, readout electronics, a battery, and a communication unit. Infectious disease testing is another significant example for nanosensor systems. There are flowstrips which are capable of testing a limited number of biochemical paramaters.

3.3.2 IoNTs for the Patient's Body

A new branch of nanotechnology combined with medicine is called nanomedicine. The more detailed definition is that nanomedicine is an interdisciplinary scientific field that involves medicine, chemistry, physics, biology, engineering, and optics. The combination of nanomedicine with IoNTs will make remarkable changes in disease prevention, diagnosis, and treatment, as well as creating a molecular communication, which is a new paradigm that uses biochemical signals to supply information exchanged among naturally and artificially created bionanosensors over short distances. It has found potential applications in targeted drug delivery and disease diagnosis/monitoring/therapy executed at the nanoscale level.

For example, nanosensors that can monitor the glucose level of the blood have been developed for the protection of diabetes patients or possible diabetes patients [43]. Furthermore, magnetic nanosensors for the detection and profiling of erythrocyte-derived microvesicles have also been used in [20, 44]. The successful implementation of such a technology will make the monitoring of individual health much easier by offering an all-time low-cost monitoring system.

Engineered nanodevices and nanostructures used in medicine and healthcare can be classified into four categories:

- first generation of nanomaterials (2000): These nanomaterials are in the class of colloids, nanostructured metals, and polymers.
- second generation of nanomaterials (2000–2005): These are the physically and chemically active structures such as actuators and amplifiers, and bioactive devices such as targeted drugs.
- third generation of nanomaterials (2005–2010): These are robotics and evolutionary biosystems.
- fourth generation of nanomaterials (2010–2020): These include the molecular devices and nanosystems that could serve as a foundation for regeneration or replacement of lost body parts.

The development of nano-based drug delivery systems includes the use of nanoparticles (Figure 3.5) in order to encapsulate the drug molecules, deliver them to the target places, and release the drug controllably to repair the damaged cells. Smart drug delivery systems are mainly used for treating cancer, neurological disorders (such as Alzheimer's and Parkinson's diseases), HIV (human immunodeficiency virus) infection and AIDS (acquired immunodeficiency syndrome) [21].

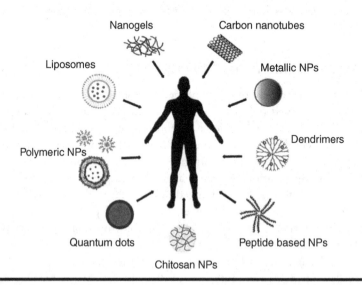

Figure 3.5 Schematic presentation of nanocarrier platforms for drug delivery systems [21].

There are many nanocarrier platforms, such as liposomes, polymersomes, quantum dots, chitosan nanoparticles, peptide-based nanoparticles, dendrimers, metallic nanoparticles, and carbon nanotubes for drug delivery used in nanomedicine (Figure 3.5) [22].

3.4 Nanosensor Fabrication and Material Science

Nanotechnology puts impact on the area of diagnostics in the health, medicine, food, environment, and agriculture sectors. By the developments and innovations in these fields, novel nanosensors and nanobiosensors have been designed and fabricated. Although sensors have a long and illustrious history, the realm of nanosensors is relatively new. A milestone chart in the development of various nanosensors from 1994 to 2005 is summarized in Figure 3.6 [23].

Nanosensors are sensing devices which convert chemical information into a quantitative useful signal with at least one of their sensing dimensions being not

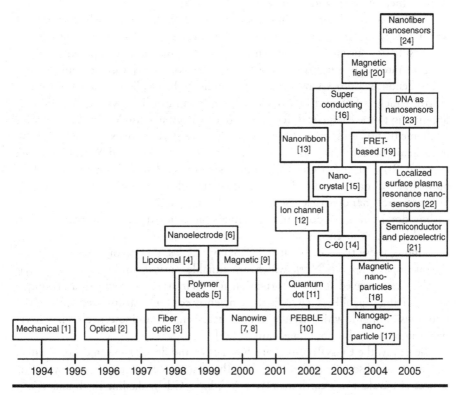

Figure 3.6 Milestone chart of various types of nanosensors [23].

greater than 100 nm. In the field of nanotechnology, nanosensors are fabricated for (a) monitoring physical and chemical phenomena in regions difficult to reach, (b) detecting biochemicals in cellular organelles, and (c) measuring nanoscopic particles in the industry and environment [23].

The various nanosensors can be grouped into three categories, and the categorization depends on the transduction mechanism for generation of output.

1. Optical nanosensors
2. Electrochemical nanosensors
3. Mechanical nanosensors

3.4.1 Optical Nanosensors

Optical biosensors rely on the detection of the change in the observed optical signal. This makes it highly compatible with various spectroscopic measurements, such as absorption, fluorescence, phosphorescence, Raman spectroscopy, surface-enhanced Raman scattering, and refraction, by detecting changes in wavelength, phase, time, intensity, and polarity of the light [24].

The first reported optical nanosensor was based on fluorescein, which is trapped within a polyacrylamide nanoparticle, and was used for pH measurement [25]. In the most basic concept, fluorescent chemosensors are molecules composed of at least one substrate binding unit(s) and photoactive component(s) [26]. The luminescence phenomenon is a process by which a fluorophore absorbs light of a certain wavelength, which is followed by emission of a quantum of light with an energy corresponding to the energetic difference between the ground and excited states [26].

Some instances, such as detection of nitrite [27], reactive oxygen species [28], pathogenic bacteria such as *S. aureus*, *V. parahemolyticus*, and *S. typhimurium*, *E. coli* [29], and detection of organophosphates [30] have been performed by optical-based nanosensors. For sensing such substrates and other substrates, luminescence nanoparticles such as gold nanoparticles, silver nanoparticles, carbon dots, quantum dots, and upconversion nanoparticles have been used in optical-based nanosensors.

Among these types of luminescent nanoparticles, rare earth doped upconversion nanoparticles (UCNPs), which are tunable optical luminescence nanomaterials, have received increased attention for their attractive features and unique upconverting capabilities. UCNPs can efficiently convert near-infrared light (NIR) into a visible light, which results in short wavelength luminescence emission via two-photon or multiphoton mechanism [22].

Generally, the UCNPs consist of a host lattice and doped lanthanide ions, which may act as an absorber and emitter ion in the host lattice material. Some crystalline lattices of trivalent rare earth ions (Sc^{3+}, Y^{3+}, La^{3+}, Gd^{3+}), alkaline earth ions (Ca^{2+}, Sr^{2+}, Ba^{2+}), or certain transition metals (Ti^{4+}, Zr^{4+}) may be used as the host materials. The most commonly used hosts are halides ($NaYF_4$, YF_3, LaF_3), oxides (Y_2O_3, ZrO_2), and oxysulfides (Y_2O_2S, La_2O_2S) [19]. On the other hand, the dopant ions located in the selected host lattice play a critical role for absorbing (e.g. Yb^{3+}) and emitting photons (Er^{3+}, Tm^{3+}, Ho^{3+}), which are responsible for the color of emitted light [31]. In order to synthesize UCNPs, a variety of chemical synthesis methods, including coprecipitation, thermal decomposition, hydro(solvo)thermal synthesis, and sol-gel, have been applied [32]. In each of these techniques, optimization of synthesis parameters in the method is crucial to obtain nanocrystals having desired size, morphology, and optical properties.

In recent years, utilizing small but efficient UCNPs, biomolecule sensors based on the upconversion fluorescence resonance energy transfer (FRET) technique have also been proven to exhibit a high sensitivity in a range of biological and chemical analyses in optical-based nanosensors [33]. A FRET system is comprised of a fluorescence donor and acceptor, which are conjugated to different biomolecular entities. The fluorescence of the donor can be effectively quenched (absorbed) by the acceptor when the distance of the donor and the acceptor comes into nanometer scale. As the distance of the donor and the acceptor is determined by their concentrations, the fluorescence intensity is linearly related to the concentration of the target when fixing either a donor or an acceptor concentration. Importantly, the FRET-UCNPs system was proved to have quite low limits of detection (LOD) in a range of experiments [34]. Wang et al. [34] firstly exploited the FRET system using UCNPs as energy donors and gold (Au) NPs as energy acceptors to detect goat antihuman immunoglobulin (IgG), reaching a low LOD of 0.88 mg/mL.

Wolfbeis and coworkers developed the FRET system, employing the acceptor of biotinylated Au-NPs and the donor of the avidin-modified $NaYF_4$:Yb^{3+}/Er^{3+} NPs in order to detect trace amounts of avidin [35]. The scheme of the FRET system in their work is shown in Figure 3.6. When Au-biotin NPs were added into the solution of avidin-conjugated UCNPs, it will bond to the surface of the UCNPs due to the sensitive and selective interaction between avidin and biotin. Since the strong absorption at ~541 nm of Au NPs well matches the UC emission of $NaYF_4$: Yb^{3+}, Er^{3+} NPs, the green emission in UCNPs will be quenched due to the FRET process. A resulting linear quenching of green upconversion emissions allows to detect and trace amounts of the avidin proteins as shown in Figure 3.7.

In another recent study, Liu et al. reported a turn-on Hg^{2+} nanosensor, which was based on the FRET between the long-strand aptamers-functioning as Up Conversion NanoParticles (UCNPs) and short-strand aptamers-functioning as Gold NanoParticles (GNPs) [36]. In the absence of Hg^{2+}, FRET between UCNPs and GNPs has occurred because of the specific matching between two aptamers, resulting in the fluorescence quenching of UCNPs. In the presence of Hg^{2+}, long-stranded aptamers have been folded back into a hairpin structure due to the stable binding interactions between Hg^{2+} and thymine, leading to the release of GNPs from UCNPs, resulting in the quenched fluorescence

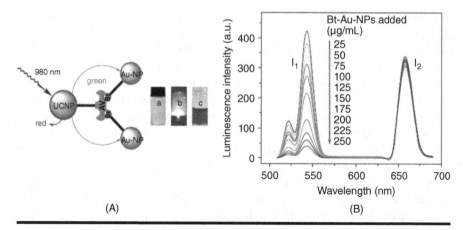

(A)   (B)

Figure 3.7 **(A) Illustration of the binding of biotinylated gold nanoparticles to avidinylated UCNPs. (a) Colorless suspension of UCNP under visible light; (b) UCNPs with green luminescence under 980-nm laser excitation; (c) adding red Au-NPs under visible light. (B) Luminescence of the UCNPs (excited at 980 nm) after addition of varying concentrations of biotinylated gold NPs [35].**

restoration (Figure 3.8). Under the optimized conditions, the nanosensor has achieved a linear detection range of 0.2-20 μM and a low detection limit (LOD) of 60 μM. Meanwhile, the nanosensor has also shown a good selectivity feature and has been applied to detect Hg^{2+} in tap water and milk samples with good precision.

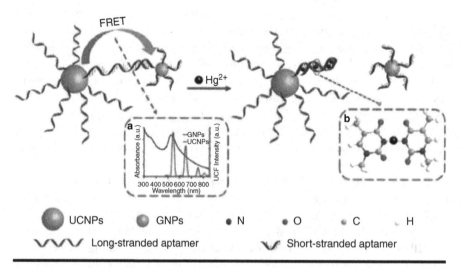

Figure 3.8 **Schematic illustration of the UCNPs-aptamers-GNPs FRET sensor for Hg^{2+} [36]. (a) The absorbance and intensity of GNPs and UCNPs complex; (b) The DNA molecules and UCNPs.**

3.4.2 Electrochemical Nanosensors

Electrochemical sensors are the most commonly used and widely accepted sensor functions on the principle of electrochemistry. The electron consumed or generated during biointeraction produces electrochemical signals, measured by electrochemical methodologies. The electrochemical nanosensors rely on chemical reactions between nanofabricated chemicals, biomolecules, and the biological element and target analyte to produce or consume ions or electrons, measured as voltage, current, or impedance. Based on their working principle, electrochemical nanosensors device could be categorized in amperometry, voltammetry, and potentiometry.

The amperometric sensor is a variant of an electrochemical sensor that continuously measures current generated due to the redox reaction of an electroactive species. The working principle behind amperometric sensors is the measurement of the current-potential relationship in an electrochemical cell where equilibrium is not established. The current is quantitatively related to the rate of the electrolytic process at the sensing electrode (also known as the working electrode) whose potential is commonly kept constant using another electrode (the so-called reference electrode) [37].

Panraksa and co-workers developed a paper-based amperometric sensor for determination of acetylcholinesterase (AChE) using a screen-printed graphene electrode [38]. AChE is an important enzyme, which is found mainly in the central and peripheral nervous system, and the abnormal function of AChE can accelerate and promote the aggregation of amyloid-betapeptides, which plays an important role in the development of Alzheimer's disease and other neurological diseases. According to the study, the amperometric detection of AChE is based on the determination of thiocholine (TCh) produced from hydrolysis of acetylthiocholine chloride (ATCh) by AChE. To detect TCh, the ATCh-immobilized sheet was stacked onto the detection sheet using double adhesive tape; then samples of AChE were dropped on the back side of an ATCh immobilized sheet with only 1 minute of incubation time (Figure 3.9). Under optimized conditions, the LOD from the experiment of AChE determination was 0.1 U/mL with AChE concentration in range of 0.1–15 U/mL [38].

The voltammetric sensor is a sensor that measures the varying current in a controlled way. Cyclic voltammetry is preferably used to get the redox potential and electrochemical reaction rates of the electrochemical reaction with analyte. The voltage parameter varies between the reference electrode and working electrode, by measuring the current between the working electrode and the counter electrode. The obtained data plotted as current versus voltage is known as a voltammogram.

Rassas and coworkers designed and fabricated a voltammetric glucose biosensor, based on the encapsulation of glucose oxidase (GOx) in a chitosan/κ-carrageenan (CHIT/CAR) polyelectrolyte complex (PEC) using a simple coacervation process [39]. Compared to biosensors based on a chitosan film, a more sensitive

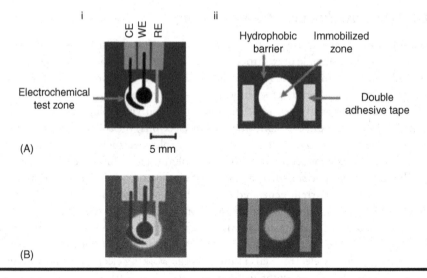

Figure 3.9 Paper-based electrochemical sensor: (A) The basic design of the developed sensor consisting of (i) the detection area with electrochemical detection (WE, working electrode; RE, reference electrode; CE, counter electrode) and (ii) ATCh immobilized sheet. (B) Image of developed sensor [38].

voltammetric detection of glucose has been obtained. Using square-wave voltammetry (SWV), the CHIT/CAR PEC-based biosensor has exhibited a wide linearity range from 5 µM to 7 mM glucose with a detection limit of 5 µM. The nanosensor has excellent selectivity against ascorbic acid, uric acid, and urea, and the applicability of the biosensor for glucose detection in spiked saliva samples has been demonstrated in the study [39].

Potentiometric sensors measure potential at the working electrode with respect to the reference electrode. The output signal is generated because of accumulation of ion at ion-selective electrodes and ion-sensitive field-effect transistors at equilibrium. Potentiometric sensors detect ions such as Na^+, K^+, Ca^{2+}, H^+, or NH_4^+ in complex biological matrices by sensing changes in electrode potential.

In a recet study, Silva and co-workers reported a label-free potentiometric immunosensor toward Salmonella Typhimurium (ST) assembled in a homemade pipette-tip electrode [40]. The signal-output amplification was implemented on a gold nanoparticle polymer inclusion membrane (AuNPs-PIM), which was used as sensing platform and for antibody immobilization. Additionally, a marker ion was used to detect the antibody-antigen binding event at the electrode surface. The immunosensor construction was performed in several steps: (i) gold salt ions' extraction in PVC membrane; (ii) AuNPs' formation using Na_2EDTA as reduction agent; and (iii) antibody anti-Salmonella conjugation on AuNPs-PIM in pipette-tip electrodes. The potential shift observed in potentiometric measurements was derived simply from the blocking effect in the ionic flux caused by antigen-antibody conjugation, without extra steps, mimetizing the ion-channel sensors. A detection limit of 6 cells mL−1 was attained. As proof of concept, recovery studies were performed in spiked commercial apple juice samples with success.

3.4.3 Mechanical Nanosensors

The earliest mechanical nanosensor was proposed by Binh and co-workers [41] for measuring the vibrational and elastic characteristics of a nanosphere attached to a tapered cantilever. This work is important for application in nanodevices components and nanoscale subassemblies in microelectronic devices. Instead of measuring the vibrational and elastic properties of the subassemblies attached to a surface, Binh and co-workers [41] introduced the concept of producing replicas of these objects from heating fine wires terminated with sharp tips. In another study, Hierold explored the possibility of down-scaling the mechanical inertia sensors from the microscale to nanoscale [42]. The sensing force is measured as a result of pressure, acceleration, and yaw rate that displaces the sensing electrode against the spring force. The change of distance with respect to the counter electrode is then measured by a change of the capacitance. Such microscale mechanical inertial sensors could be scaled down into nanosensors provided that self-assembly of nanostructures becomes a well-controlled fabrication technology.

3.5 Nanosensor and Energy Harvesting

Typically, batteries are used to supply power in IoNT communication networks. However, their finite capacity and replacement necessity are significant problems, especially in health applications [43]. Increasing the battery size/capacity is not a solution as the cost and weight of the nanosensor will simultaneously increase. An increase in the nanosensor weight can create a bulky system and hurt the desired mobility feature in IoNT. Therefore, recent studies demand the use of energy harvesting (EH) methods instead of using these batteries. EH in nanosensors and nanotechnology in general is a relatively recent trend toward prolonging the IoNT system lifetime. By employing state-of-the-art ambient energy harvesting circuits, self-powered nanosensors have been introduced to utilize various ambient energy sources such as the body heat and movement, radio frequencies (RF), and/or solar energy [43–45]. Body heat has been proven to be a suitable candidate for energy harvesting in order to fully eliminate the use of batteries in IoNT [46, 47]. The study presented in [51] focuses on EH methods, and it briefly explains the working principles of each method. In [51], EH methods are classified by considering the sensor application specifications. Similarly, [52] presents the classification of the EH methods specifically for sensor networks. Another study presented in [53] introduces EH methods for body sensor networks.

In general, solar, thermal, mechanical movement or vibration, and ambient RF sources can be used to generate electrical energy [48]. The vibration or movement-based energy scavenging can be divided into three major categories: electrostatic, electromagnetic, and piezoelectric [47–51]. Thus, the availability of one of these sources is of utmost importance for IoNT applications. Furthermore, the energy extraction capacity, physical size, robustness, and output impedance characteristics of the harvesting circuits play an important role in prolonging the nanonetwork lifetime.

3.6 Nanosensors and Body Networks

A wireless body area network (WBAN) is a network that provides a continuous monitoring over or inside the human body for a long period. It usually consists of the following components in the IoNT paradigm:

- Nanonodes: These are the nanosensor devices. Due to the limited energy, limited memory, and reduced communication capabilities, they can only perform simple computation tasks and can transmit over very short distances.
- Nanorouters: These are the nanodevices with slightly larger computational resources than nanonodes and can aggregate information from limited nanonodes and also can control the behavior of nanonodes. However, their size is bigger, and thus, their deployment is more invasive.
- Nano-micro interface: They are used to aggregate the information forwarded by nanorouters and send the information to the microscale devices. At the same time, they can send the information from microscale to nanoscale. Nano-micro interfaces are hybrid devices able not only to communicate in the nanoscale using the nanocommunication techniques but also can use classical communication paradigms in micro/macro communication networks.
- Gateway: It enables the users to control/monitor the entire system remotely over the Internet. It is a hybrid device not only able to communicate in the nanoscale, but also can use classical communication paradigms in micro/macro communication networks such as Wifi, Bluetooth, and Zigbee.

The nanocommunication in WBAN can utilize one of the following technologies: nanomechanical, acoustic, electromagnetic, and chemical or molecular communications. It supports transmissions of real-time traffic such as data, to observe the status of vital organ functionalities. The WBAN has recently gained attraction for its invaluable capability in twenty-four-hour health monitoring to detect chronic diseases as well as "fatal event" prognostics such as a heart attack. It also finds its application in sports to provide quantitative and qualitative information from athletes while playing sports.

One of the main areas that the WBAN has been used in is the area of healthcare. Where systems for monitoring the internal well-being of the human body have been developed, these systems usually employ a vast number of nanosensors embedded in the human body. They continuously communicate with each other and with the outside environment, forming a network known as the WBAN. Mainly due to their tiny sizes, WBANs introduce difficulties in both hardware and software design. Detailed explanations on the hardware design of the electromagnetic wireless nanosensors can be found in [54]. Especially for the software part, the communication layer stack needs fine tuning as the hardware imposes many restrictions. The physical signaling is at the THz levels, due to antenna size, requiring special modulation techniques. On the other hand, promising research is being done by using graphene-based plasmonic materials for antennas to overcome signaling difficulties [55–58].

The majority of the attempts in literature have focused mostly on the nanosensor architecture and given a taxonomy for energy conservation schemes. It involves three main classes of energy conservation schemes, namely the duty-cycling, data-driven, and mobility-based schemes. Duty cycling is an effective operating system (OS) and Hardware (HW)-level optimization technique, in which the components can cycle between various levels of "SLEEP" and "WAKEUP" states. The idea is to put the components into "SLEEP" state when they are not needed and change the state to "WAKEUP" when they are needed. The data-driven schemes build a statistical model of the sensed data for reducing the actual sensing activities with the prediction. The idea in mobility-based schemes is to equip some or all nodes with mobility when the energy cost of mobility can be used to further reduce the communication energy cost. Clustering is one of the optimization techniques that is offered for efficient energy usage (both network level and nanosensor level) and for data aggregation" (prevents data duplication). While the topology of the WBAN is clustered logically, introducing hierarchy, the transmission flow is also controlled by elected cluster heads (CH). Various algorithms are proposed for the election process of the CHs, which is a crucial process for energy optimization. The main idea is balancing the energy usage of the clusters and increasing the lifetime of the network. The next phase is the cluster formation, in which nodes based on some heuristics decide to join to a certain CH for forming a cluster.

Unfortunately, current networking protocols cannot be applied to nanocommunication. They are too complex for nanosensors, and the energy constraints of nanosensors cannot sustain such protocols. Due to the nanoscale dimensions of the sensor nodes and the size of their antennas, THz frequency is the only band where the communication can be done. As a result, new protocols, new modulation techniques, and new signal-encoding methods are necessary for THz frequency communications.

3.7 Open Research Issues

Environment Monitoring: IoNT helps in controlling and enhancing the productivity in agriculture by monitoring the temperature, humidity, and pH of the soil. Several types of nanoparticles and their composites have been found useful in pesticide management and antiphytopathogens. Nanoparticles are also used to control fungi in plants [7]. These particles can be embedded into a nanosensor and the deployment of utilization of these particles can be controlled and monitored by IoNT. The fertilizer can be released to the plant on an on-demand basis using IoNT. A drone can be used to monitor the health of the plants by studying the air around them. If any information has been found about any pest/fungi, the drone can release the nanosensor on the infected plant. Using IoNT, the information about the infected plant can be sent to the owner and he/she can trigger the release of nanosensors.

Detection of Viruses and Bacteria: Due to rapid changes in the weather conditions, there are various environmental side effects where bacteria or viral infections significantly appear. The nanosensors that are deployed in various locations in the environment (inside and outside the buildings) can be able to identify and take the necessary measures that can be used to reduce the infection and bacteria. For instance, for smart cities, smart nanosensors can be produced in order to keep a city clean. The nanosensor can be placed near the garbage bin and once any pollutant is found in the air a trigger is generated to release cleaning nanosensors to clean up that particular area.

Food Packing: Nanosensors can provide real-time status of the food freshness, which helps in obtaining the exact expiration date. Nanosensors can be used for tracking the product, brand protection, and authentication of documents like passports. Once these nanosensors are equipped with the facility to communicate wirelessly using IoNT, they may replace RFIDs and barcodes. The packaging can be made interactive, where the item will send a message to the personal device if it is not stored or placed in proper condition. To avoid waste, a message can be send to the mobile device in range indicating its expiration date. Also in the delivery chain, the product can be tracked on real time without any involvement of the courier service.

Nanosensors for Avoiding Traffic Accidents: Nanosensors are used to identify the drowsiness of a driver and transfer the information to a device using Bluetooth. IoNT increases the monitoring range when compared to Bluetooth. It should be added that every vehicle should be equipped with a sensor to detect drowsiness/drunkenness and transfer the information to the central control room. Further advancement can be made by deploying and networking nanosensors within the vehicle to help in deactivating the engine/vehicle when the condition of the driver is not appropriate to drive. IoNT enables assisted driving where the traffic conditions are routed to the navigation system present in the vehicle and will help in choosing the optimized path. The entire navigation within a city will available at the central monitoring station and make monitoring of important activities easy.

Defense: Nanotechnology research remains a major national initiative in most nations. It addresses the needs and challenges in national security and public safety. The significant breakthrough in nanotechnology can have a high potential to affect military capabilities. The key capabilities include power and energy-efficient materials, highly resistive materials, and coatings for platforms and weapons such as nanoarmors, nanomedic future warfighting, and chemical/biological warfare defense machines.

Energy Harvesting (EH) methods in nanosensors can be classified based on the ambient energy sources from which they generate electrical energy. That is because of the fact that utilization of the harvesters depends highly on the availability of the ambient sources. The output voltages, output power densities, sizes, and ideal conditions for each harvesting mode have to be considered as well. Nanoscale batteries cannot store much energy for the long duration of the non-stopping operations in the nanosensors. Although supercapacitors are used nowadays to replace the batteries' usage due to their small size, EH in nanosensors aims to completely remove the energy storage element from the nanosensor system and/or IoNT paradigm. Therefore, removing the energy storage units while utilizing efficient energy harvesting methods can be still a hot research topic.

Meanwhile, there are several challenges which face the implementation of the nanosensor communication networks (i.e., the WBAN) using nanotechnology. For example, security of transmitted data and privacy concerns is a great issue in this domain. Moreover, modeling the wireless communication channel over/through the human body is another significant research area nowadays. Unfortunately, current wireless network protocols cannot be applied to nanocommunication. They are too complex for nanosensors, and the energy constraints of nanosensors cannot sustain such protocols.

3.8 Concluding Remarks

The development of nanotechnologies, nanosensors, and IoNTs will have a great impact on advanced development in almost every field in near future, and they will lead to development of inexpensive, portable devices for the broad detection, identification, and quantification of biological and chemical substances. Researchers are currently working in development of nanosensors comprising IoNT for live deployment in varied areas in the near future. In this paper, in-depth review with regard to nanosensors for IoNTs is presented, which is regarded as next evolutionary step in the world of nanotechnology, in addition to nanosensors, applications, and research areas.

References

1. Murty B.S., Shankar P., Raj B., Rath B.B., Murday J. 2013. *The textbook of nanoscience and nanotechnology.* New York: Springer.
2. Zijlstra P., Paulo P.M.R., Orrit M. 2012. Optical detection of single non-absorbing molecules using the surface plasmon resonance of a gold nanorod. *Nat. Nano.*, 7, 379–384.
3. Srivastava A.K., Dev A., Karmakar S. 2018. *Nanosensors and nanobiosensors in food and agriculture. Envir. Chem. Lett.*, 16, 161–182.
4. Nezami A., Dehghani S., Nosrati R., Eskandari N., Taghdisi S.M., Karimi G. 2018. Nanomaterial based biosensor and immunosensors for quantitative determination of cardiac troponins. *J. Pharm and Biomed. Analy.*, 159, 425–536.
5. Lee S.J., Jung C., Choi K., Kim S. 2015. Design of wireless nanosensor networks for intrabody application. *Int. J. Dist. Sens. Netw.*, 176761.
6. Sitti M., Ceylan H., Hu W., Giltinan J., Turan M., Yim S., Diller E. 2015. Biomedical applications of untethered mobile milli/microrobots. *Proc. IEEE Inst. Electr. Electron Eng.*, 103, 205–224.
7. Limaye V., Fortwengel G., Limaye D. 2014. Regulatory roadmap for nanotechnology based medicines. *Int. J. Drug Regul. Aff.*, 2(4), 33–41.
8. Al-Turjman F. 2018. "QoS–aware data delivery framework for safety-inspired multimedia in integrated vehicular-IoT", *Elsevier Computer Communications Journal*, 121, 33–43.
9. Atzori L., Iera A., Morabito G. 2010. The Internet of Things: A survey. *Elsevier Computer Networks*, 54, 2787–2805.
10. Darianian M., Michael M.P. 2008. Smart home mobile RFID-based Internet-of-things systems and services. *Advanced Computer Theory and Engineering, ICACTE '08. International Conference*, 116–120.
11. Al-Turjman F. 2018. The road towards plant phenotyping via WSNs: An overview, *Elsevier Computers & Electronics in Agriculture.* DOI: 10.1016/j.compag.2018.09.018.
12. Amendola S., Lodato R., Manzari S., Occhiuzzi C., Marrocco G. 2014. RFID technology for IoT-based personal healthcare in smart spaces. *IEEE Internet of Things Journal*, 1, 144–152.
13. Manzoor A. 2016. RFID in health care–building smart hospitals for quality healthcare. *Int. J. User-Driven Healthcare*, 6, 21–45.
14. Al-Turjman F. 2019. 5G-enabled devices and smart-spaces in social-IoT: An overview, *Elsevier Future Generation Computer Systems*, 92(1), 732–744.
15. Gaynor M., Waterman J. 2016. Design framework for sensors and RFID tags with healthcare applications. *Health Policy Technol.*, 5, 357–369.
16. D'Andrea A., Ferri F., Grifoni P. 2016. Multimodal social networking for healthcare professionals. *Int. J. Comput. Clin. Practice*, 1, 15–27.
17. Premkumar M., Girish M., Karthikiran I. 2016. Smart errands and automatic billing reckoning system using LIFI and OTG android technology. *Automation and Autonomous System*, 8, 198–201.
18. Kassal P., Steinberg I.M., Steinberg M.D. 2013. Wireless smart tag with potentiometric input for ultra low power chemical sensing. *Sens. Actuators, B*, 184, 254–259.
19. Hoss U., Budiman E.S. 2017. Factory-calibrated continuous glucose sensors: The science behind the technology. *Diabetes Technol. Ther.*, 19, S44–S50.
20. Rho J., Chung J., Im H., Liong M., Shao H., Castro C.M., Weissleder R., Lee H. 2013. Magnetic nanosensor for detection and profiling of erythrocyte-derived microvesicles. *ACS Nano.*, 7, 11227–11233.

21. Liu D., Yang F., Xiong F., Gu N. 2016. The smart drug delivery system and its clinical potential. *Theranostics*, 6, 1306–1323.
22. Buddola A.L., Kim S. 2018. Recent insights into the development of nucleic acid-based nanoparticles for tumor targeted drug delivery. *Coll. Surface B: Bioint.*, 172, 315–322.
23. Lim T.C., Ramakrishna S. 2014. A conceptual review of nanosensors. *Zeitschrift für Naturforschung A*, 61, 402–412.
24. Ahuj D., Parande D. 2012. Optical sensors and their applications. *J. Sci. Res. and Rev.*, 1, 060–068.
25. Sasaki K., Shi Z.Y., Kopelman R., Masuhara H. 1996. Three-dimensional pH micro-probing with an optically-manipulated fluorescent particle. *Chem. Lett.*, 25, 141.
26. Cusano A., Arregui F.J., Giordano M., Cutolo A. 2012. *Optochemical nanosensors.* Boca Raton, FL: CRC Press.
27. Xiang G., Wang Y., Zhang H., Fan H., Fan L., He L., Jiang X., Zhao W. 2018. Carbon dots based dual-emission silica nanoparticles as ratiometric fluorescent probe for nitrite determination in food samples. *Food Chem.*, 260, 13–18.
28. Gong Y.J., Lv M.K., Zhang M.L., Kong Z.Z., Mao G.J. 2019. A novel two-photon fluorescent probe with long-wavelength emission for monitoring HClO in living cells and tissues. *Talanta*, 192, 128–134.
29. Liu Y., Wei Y., Cao Y., Zhu D., Ma W., Yu Y., Guo M. 2018. Ultrasensitive electrochemiluminescence detection of Staphylococcus aureus via enzyme-free branched DNA signal amplification probe. *Biosensors and Bioelect.*, 117, 830–837.
30. Arjmand M., Saghafifar H., Alijanianzadeh M., Soltanolkotabi M. 2017. A sensitive tapered-fiber optic biosensor for the label-free detection of organophosphate pesticides. *Sens. Act. B: Chem.*, 249, 523–532.
31. Wang F., Banerjee D., Liu Y., Chen X., Liu X. 2010. Upconversion nanoparticles in biological labeling, imaging, and therapy. *Analyst*, 135, 1839–1854.
32. Wen S., Zhou J., Zheng K., Bednarkiewicz A., Liu X., Jin D. 2018. Advances in highly doped upconversion nanoparticles. *Nature Commun.*, 9, 2415–2427.
33. Chen G., Song F., Xiong X., Peng X. 2013. Fluorescent nanosensor based on fluorescence resonance energy transfer (FRET). *Ind. Eng. Chem. Res.* 2013, 52, 11228–11245.
34. Wang M., Hou W., Mi C.C., Wang W.X., Xu Z.R., Teng H.H., Mao C.B., Xu S.K. 2009. Immunoassay of goat antihuman immunoglobulin G antibody based on luminescence resonance energy transfer between near-infrared responsive NaYF$_4$:Yb, Er upconversion fluorescent nanoparticles and gold nanoparticles. *Anal. Chem.*, 81, 8783–8789.
35. Saleh S.M., Ali R., Hirsch T., Wolfbeis O.S. 2011. Detection of biotin–avidin affinity binding by exploiting a self-referenced system composed of upconverting luminescent nanoparticles and gold nanoparticles. *J Nanopart Res.*, 13, 4603–4611.
36. Liu Y., Ouyang Q., Li H., Chen M., Zhang Z.Z., Chen Q. 2018. A turn-on fluorescence sensor for Hg in food based on FRET between aptamers fuctionalized upconversion nanoparticles and gold nanoparticles. *J. Agric. Food. Chem.* 66, 6188–6195.
37. Trnkova L., Adam V., Hubalek J., Babula P., Kizek R. 2008. Amperometric sensor for detection of chloride ions. *Sensors (Basel)*, 8, 5619–5639.
38. Panraksa Y., Siangproh W., Khampieng T., Chailapakul O, Apilux A. 2018. Paper-based amperometric sensor for determination of acetylcholinesterase using screen-printed graphene electrode. *Talanta*, 178, 1017–1023.

39. Rassas I., Braiek M., Bonhomme A., Bessuelle F., Rafin G., Majdoub H., Jaffrezic-Renault N. 2019. Voltammetric glucose biosensor based on glucose oxidase encapsulation in a chitosan-kappa-carrageenan polyelectrolyte complex. *Mater. Sci. Eng: C*, 95, 152–159.

40. Silva N.F.D., Magalhaes J.M.C.S., Oliva-Teles T., Freire C., Delerue-Matos C. 2019. In situ formation of gold nanoparticles in polymer inclusion membrane: Application as platform in a label-free potentiometric immunosensor for salmonella typhimurium detection. *Talanta*, 194, 134–142.

41. Binh V.T., Garcia N., Lavanuyk A.L. 1995. Estimations for the characteristics of GHz range nanocantilevers: Eigenfrequencies and quality factors. *Surface Sci.*, 328, 307–342.

42. Al-Turjman F. 2017. Cognitive-node architecture and a deployment strategy for the future sensor networks, *Springer Mobile Networks and Applications*. DOI: 10.1007/s11036-017-0891-0.

43. Demir S., Al-Turjman F. 2018. Energy scavenging methods for WBAN applications: A review. *IEEE Sensors Journal*, 18(16), 6477–6488.

44. Al-Turjman F. 2017. A rational data delivery framework for disaster-inspired Internet of Nano-Things (IoNT) in practice. Springer Cluster Computing, DOI: 10.1007/s10586–017–1357–7.

45. Al-Turjman F., Imran M., Vasilakos A. 2017. Value-based caching in information-centric wireless body area networks. *Sensors Journal*, 17(1), 1–19.

46. Al-Turjman F., Imran M., Vasilakos A. 2017. Value-based caching in information-centric wireless body area networks, *Sensors Journal*, 17(1), 1–19.

47. Al-Turjman F. 2017. A cognitive routing protocol for bio-inspired networking in the Internet of Nano-Things (IoNT). *Springer Mobile Networks and Applications*, DOI: 10.1007/s11036-017-0940-8.

48. Rashidzadeh H., Kasargod P.S., Supon T.M., Rashidzadeh R., Ahmadi M. May 2016. Energy harvesting for IoT sensors utilizing MEMS technology. *2016 IEEE Canadian Conference on Electrical and Computer Engineering* (CCECE 2016), 1–4.

49. Lu C., Raghunathan V., Roy K. 2011. Efficient design of micro-scale energy harvesting systems. *IEEE J. Emerging Sel. Topics Circuits Syst.*, 1(3), 254–266.

50. Al-Turjman F., Hassanein H., Ibnkahla M. 2013. Efficient deployment of wireless sensor networks targeting environment monitoring applications, *Elsevier: Computer Communications Journal*, 36(2), 135–148.

51. Zahid Kausar A.S.M., Reza A.W., Saleh M.U., Ramiah H. October 2014. Energizing wireless sensor networks by energy harvesting systems: Scopes, challenges and approaches. *Renewable and Sustainable Energy Reviews*, 38, 973–989.

52. KarimShaikh F., Zeadally S. March 2016. Energy harvesting in wireless sensor networks: A comprehensive review. *Renewable and Sustainable Energy Reviews*, 55, 1041–1054.

53. Al-Turjman F. 2017. Information-centric sensor networks for cognitive IoT: an overview, *Annals of Telecommunications*, 72(1), 3–18.

54. Akyildiz I.F., Jornet J.M. 2010. Electromagnetic wireless nanosensor networks. *Nano Comm. Networks*, 1(1):3–19.

55. Jornet J.M. December 2013. Fundamentals of electromagnetic nanonetworks in the terahertz band. PhD thesis, School of Electrical and Computer Engineering, Georgia Institute of Technology, Atlanta, GA.

56. Jornet J.M., Akyildiz I.F. 2010. Graphene-based nano-antennas for electromagnetic nanocommunications in the terahertz band. In *Proceedings of the 4th European Conference on Antennas and Propagation*, 1–5.

57. Jornet J.M., Akyildiz I.F. 2013. Graphene-based plasmonic nano-antenna for terahertz band communication in nanonetworks. *IEEE Journal on Selected Areas in Comm.*, 31(12):685–694.

58. Hasan M.Z., Al-Rizzo H., Al-Turjman F. 2017. A survey on multipath routing protocols for QoS assurances in real-time multimedia wireless sensor networks. *IEEE Communications Surveys and Tutorials*, 19(3), 1424–1456.

Chapter 4

mm-Waves in the Internet of Nano-Things

Fadi Al-Turjman[1] and Jehad Hamamreh[2]

[1]Department of Computer Engineering, Antalya Bilim University, Antalya, Turkey
[2]Antalya Bilim University, Antalya, Turkey

Contents

4.1 Introduction...45
4.2 mm-Wave Propagation Models...47
4.3 Antenna Design Aspects 49
4.4 Confidentiality Aspects 51
4.5 Open Research Issues..54
4.6 Conclusion...54
References 55

4.1 Introduction

Millimeter and terahertz waves are both electromagnetic waves just like microwaves, radio waves, and visible and infrared light [1, 2]. Electromagnetic waves, up to and including microwaves that have relatively low frequencies, are generally unaffected by rain and other atmospheric effects. This has made them suitable for long-range radio communications such as television (TV) and radio broadcasting communications. Nevertheless, as the wavelength becomes shorter, it becomes more difficult for those waves to penetrate through objects/medium molecules. Millimeter and terahertz waves suffer from attenuation caused by rain and resonant absorption in oxygen and water molecules, so they are usually unsuitable for long-range

45

radio communications [3, 4]. However, with the advances of technologies and low-cost integration solutions, millimeter technology has started to gain a great deal of momentum from academia, industry, and standardization bodies [5]. Particularly, advanced beamforming technologies enabled by the utilization of large numbers of millimeter wave antennas with small size dipoles can promote the dense antenna arrays in compact base stations for small cell network densification in congested urban areas to increase directionality, overcome propagation path loss, and expand data capacity while connected to the cloud. The millimeter-Waves (mm-Waves) and/or terahertz (THz) systems have been used in the medical industry via nanodevices and sensors, in order to transmit patients' vital signs (e.g. SPO2, pulse, blood pressure, etc.) at short ranges to reduce the number of leads from the patient's body. Moreover, recent development of radio frequency integrated circuit (RFIC) design and relatively inexpensive power efficient complementary metal oxide semiconductor (CMOS) processing for semiconductor manufacturing have opened the millimeter wave bands and sub-millimeter wave bands for commercial use, as evidenced by the IEEE 802.15.3c and IEEE 802.11ad [12–16].

The vision of millimeter wave communications is to unleash the 30–300 GHz spectrum with the potential of a new spectrum suitable for wearable devices and its IoNT applications. That is mainly due to these communications' wavelengths in the range of millimeter order [6–9]. Short wavelengths have been proven to be a great alternative for large amounts of data transmissions at once in mm-Waves. The mm-Waves' vision enables low cost backhauls, last-mile wireless broadband access, low interference in highly dense millimeter small cells, low latency in uncompressed high-definition multimedia transfers, and the wireless access to the cloud [11]. This can dramatically change the future of wearable devices as it can affect its inter- and/or intra-communications techniques. And it makes these devices widely accessible through the Internet and other advanced communication technologies such as the fifth generation (5G) in cellular networks.

In this chapter, we focus on design aspects of the mm-Waves and their applications in the Internet of Nano-Things (IoNT) era. The IoNT is the interconnection between nanoscale devices with the existing communicating network technology and/or the Internet. The intrabody network is one of the architectures of the Internet of Nano-things, where nanomachines, such as nanosensors injected into the human body, are controlled on a macroscale level by the healthcare service provider over the Internet. Authors in [9] provide an example of a wearable device with a micro-gateway, which receives data from nanosensors in the human body and transmits the data to the patient's doctor. The interconnected office is another architecture. In this one, all devices in an office are fitted with nanotransmitters which allow them to connect to the Internet, and hence, the user is able to effortlessly locate and know the condition of the office elements. In IoNT, micro- and nanodevices are connected to each other and the Internet via local gateways applying the mm-Wave technology. Integrating IoNT with other local area networks using this mm-Wave technology can significantly expand the range of services which can be provided to public users as well as decision makers. Due to the peculiarities of the mm-Wave/THz communication, which enables enormous bandwidth and negligible transmission delay, the probability of collision is considerably less in IoNT than that of traditional wireless networks. Furthermore, the mm-Wave technology has been claimed to provide a unique and secure identification method for the connected physical items, in addition to managing the rapidly changing data in a real-time fashion. And thus, IoNT has applied it in/out door environments for healthcare and industrial applications, which makes it possible to connect everything at any time and any place. However, the unique characteristics of the mm-Wave/THz communication band and specific challenges in IoNT, such as energy constraints and random network topology, prevent the direct implementation of traditional wireless communication techniques without a careful study beforehand [10].

In this chapter, we highlight the technical issues that can affect its applicability and practicality in critical missions. Accordingly, the mm-Wave signal propagation models are investigated and analyzed in this chapter for better understanding of the physical properties and the unique characteristics in this communication technology. Moreover, the antenna design aspects and security issues are covered. Finally, we wrap up this chapter by highlighting the open research issues and a few concluding remarks in this area.

4.2 mm-Wave Propagation Models

To get an insight on the challenge associated with mm-Wave propagation, we need to revisit and have a look at Friis' Law, which represents the most fundamental, basic propagation model applicable in free-space environments. In this model, the transmitted power, P_t, and received power, P_r, in linear scales are related by the following formula [17]:

$$\frac{P_r}{P_t} = G_r G_t \left(\frac{\lambda}{4\pi d} \right)^2, \ \lambda \times f = c \tag{4.1}$$

where d in Eq. (4.1) is the separation distance between the transmitter and receiver, λ is the wavelength, f is the frequency of the wave, c is the speed of light, G_t is the transmit antenna gain and G_r is the receive antenna gain. Friis' Law indicates that the isotropic path loss (i.e., the degradation in the transmitted power) increases with the increment in the squared wave frequency, f^2. This fact implies that mm-Wave propagation will experience a much more severe path loss compared to waves with lower frequencies (i.e., below 6 GHz, which are traditionally used in cellular networks such as 2G-GSM, 3G-UMTS, and 4G-LTE).

However, the challenge of having severe propagation path loss in mm-Waves can surprisingly be solved by using highly directional array antennas. To understand how this works, we must pay attention to the fact that for a given physical antenna aperture, the maximum directional gains generally scale with frequency as G_r, G_t $\propto f^2$, since more antenna elements can be squeezed and fit into the same physical area as a result of having a small wavelength in the mm-Wave range. Therefore, the directional gain, which can be resulted from the ability to use massive antenna elements to form the transmission beam, can easily compensate and substitute the increased propagation path loss at mm-Wave frequencies [18]. From this intuitive explanation, one can conclude that there is a formidable need to use directional transmissions with high-dimensional antenna arrays to compensate for path loss in the mm-Wave propagation environments. This phenomenon explains why multiple-input multiple-output (MIMO) with beamforming is a defining, indispensable design component for mm-Wave communication systems [18].

Although propagation loss can roughly be predicted by Friis' formula, the measurement of actual path loss in general environments depends highly on the specific 3D location of objects and materials that can attenuate, scatter, reflect, and diffract signals. To calculate the path loss in mm-Wave bands, the ray tracing method has been used in the literature in predicting geometry-based and site-specific mm-Wave propagation, particularly in indoor environments [19]. Besides, there is a large amount of effort in developing and producing mm-Wave statistical models that characterize the distribution and statistics of path losses over an ensemble of environments [20]. Particularly, many studies have been conducted in short-distance links in indoor wireless LAN or PAN systems, which are mostly very suitable for IoNT applications. Without loss of generality, the average path loss (excluding small-scale fading) in mm-Wave environments can be described by the most common statistical model given as follows:

$$PL(d)[dB] = \alpha + 10 \times \beta \times \log_{10}(d) + \xi, \ \xi \sim N(0, \sigma^2) \tag{4.2}$$

where d is the distance between sender and receiver, α and β are environment-dependent parameters, and ξ is a lognormal term, including the variance in shadowing effects. The precise values of the parameters given in the previous model can be found in [21], [22], and [23] for short-range and indoor settings. Meanwhile, the proposed path loss formula in [55] is at THz, and has two parts: the absorption path loss and the spread path loss. Four different power spectral densities (p.s.d.) were studied by authors in [55] i.e., optimal p.s.d., flat p.s.d., the Gaussian pulse and the p.s.d. for the case of the transmission window at 350 GHz, which concluded that for the very short communication range, quite high transmission bit-rates can be supported, up to terabits per second, indicating the promising future of the nanocommunication.

Despite the ability to overcome and compensate for the challenge of propagation path loss for mm-Wave frequencies by using directional transmissions enabled by array antennas, a more significant challenge is their severe sensitivity to blockage. Different materials can attenuate mm-Wave frequencies by different levels. For instance, brick can decrease the power of mm-Wave signals by as much as 40 to 80 dB [24] whereas the human body can cause a 20 to 35 dB loss [25]. Similarly, foliage loss can also be very high depending on the type of plant leaves that the signal tries to penetrate [26]. The human body and materials used in most buildings are relatively reflective in mm-Wave bands. The scatterings and reflections from these materials enable coverage via non-line of site (NLOS) paths. This interesting feature related to NLOS coverage in mm-Waves is deemed as good news for cellular systems supporting IoNT devices since coverage in this case is possible up to 200 m from a certain transmitter [27].

Blocking models can be quantified statistically as in [28] or derived analytically from random shape theory as can be found in [29] or from geographic information as in [30]. Using these kinds of models, it is possible to analytically investigate and quantify both coverage and capacity in mm-Wave cellular networks using the theory of stochastic geometry [31]. One final notice to mention is that the standard multipath MIMO models used in lower frequencies can be directly applied to describe the spatial properties and multi-path channel characteristics of mm-Wave MIMO systems [32].

4.3 Antenna Design Aspects

Wireless communications in mm-Wave bands (28, 38, 60, 73 GHz, and beyond) has recently drawn and attracted an increasing interest due to the scarce availability of unoccupied spectrum in lower-frequency bands. One of the well-recognized advantages of mm-Wave communications is the ability to utilize electrically large antenna arrays to form highly directive beams that can combat the propagation attenuation (losses) and blockage challenges. However, the cost and power consumption of mm-Wave mixed signal analog/digital components can make it impractical and inefficient to utilize a radio frequency (RF) chain for each of the antenna elements in a large array to enable MIMO signal processing in the baseband. Hence, different mm-Wave architectures and designs that are capable of addressing the challenges of both cost and power have the potential for maximizing the benefits of mm-Wave spectrum in the next generation of wireless communication systems. These prospective antenna architectures can be divided and classified into three main beamforming antenna designs, which are discussed as follows:

▪ **Analog beamforming:** Traditional phased array (TA) architecture is a natural choice for alleviating the challenges of accommodating power-hungry RF chains and analog-to-digital/digital-to-analog converters (ADC/DAC) for each antenna element. In this architecture, beamforming is accomplished by including a variable phase shifter per antenna element as shown in Figure 4.1. Phase shifters are typically followed by transmit/receive amplification and switch (A&S) stages to meet radiated/received power requirements. Due to the

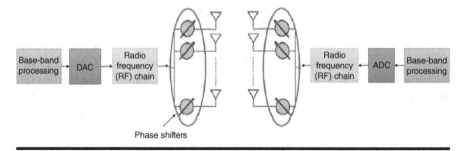

Figure 4.1   Analog beamforming architecture at the transmitter and receiver sides.

significant reduction in RF chain and ADC/DAC units count, the power consumed by the phase shifters of the large format traditional arrays becomes a critical power efficiency consideration. Larger antenna arrays are also expected to need phase shifters with more bit states and lower quantization errors to achieve desired beamforming performance, further triggering the power efficiency issue. Most importantly, the data rate of the traditional phased-array architecture is ultimately limited due to the inability to support simultaneous multiple MIMO stream transmissions.

■ **Hybrid (analog/digital) beamforming:** The need to provide multiple MIMO stream transmissions to support multiple users while maintaining a reduced number of RF chains has resulted in the introduction of hybrid (analog/digital) MIMO architectures [33,34]. The usual hybrid MIMO architecture is based on the traditional phased-array approach. Here, the RF signal from each RF chain (a total number of L) is split into antenna elements (a total number of N) and passed through phase shifters before the antenna interface as depicted in Figure 4.2. Hence, the total number of phase shifters ($NPS = N * L$) and associated power consumption dramatically increase. This usual hybrid MIMO architecture is essentially a superposition of L phased arrays—each RF chain having the capability to form and steer the narrowest possible beam with the aid of its own phase shifter network

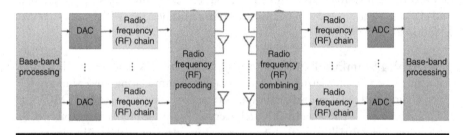

Figure 4.2   Hybrid (analog/digital) beamforming architecture at the transmitter and receiver sides.

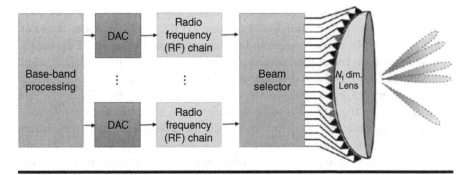

Figure 4.3   Lens array-aided beamforming architecture at the transmitter side.

(i.e., RF precoder/combiner). It should be mentioned that the high power consumption and complexity of hybrid MIMO architectures motivated the consideration of alternative architectures that exhibit reduced number of phase shifters or replace phase shifters with significantly more efficient switches.

■ **Lens array-aided beamforming:** This unique architecture, which is also named continuous aperture phased (CAP)-MIMO [35, 36], enables multibeam steering and data multiplexing with minimal complexity among all the available ones in the literature. This is achieved by concurrently generating P number of beams and multiplexing data into them by using the lens array-based antenna architecture as shown in Figure 4.3. Different beams directed toward different directions are produced by exciting a certain number of feed antennas. Data multiplexing into the feeds is attained by using a mm-Wave beam selector which represents the most complex part of the CAP-MIMO transceiver.

4.4 Confidentiality Aspects

Confidentiality, authentication, integrity, and availability are indispensable security services for IoNT communication systems. Among all these different services, confidentiality represents the most critical one, as it prevents the information leakage to eavesdroppers and unauthorized devices from one side, and constitutes a first-step protection against other common security attacks from another side. Particularly, the eavesdropped data can be exploited to cause different types of security breaches such as denial of service (DoS), identity-based attacks, man-in-the-middle attacks, session hijacking, spoofing, sniffing, and data modification. To achieve confidentiality, encryption-based approaches can be used as a classical solution against eavesdropping. However, conventional approaches based on encryption and cryptography entail many hurdles and practical difficulties for future wireless systems. As a potential alternative to cryptography, physical layer security (PLS) has emerged as a new concept that can supplement and may even completely substitute encryption-based approaches [37]. The basic, essential idea of PLS is to exploit the characteristics and features of the wireless channel, including its randomness, dependence on transceivers' locations, reciprocity in time division duplexing (TDD) system, and impairments, including noise, fading, dispersion, interference, etc. to guarantee security against eavesdropping [38]. Thus, the primary design objective of PLS is to boost and increase the performance difference between the legitimate receiver's link and that of the eavesdropper by the utilization of well-designed transmission schemes.

The adoption and application of PLS to mm-Wave communication systems, which are deemed as suitable candidates for IoNT wearable devices, is a noticeably emerging area of research. This is motivated and supported by the fact that the features and characteristics of the mm-Wave channel are significantly different from those of microwave channels (u-Wave). Accordingly, the unique and special characteristics of mm-Wave channels, including very short-wave length, larger system bandwidth, directionality by using massive antenna arrays, and short-range transmissions due to severe path-loss propagation and sensitivity to blockage, can all be judiciously exploited together to help improve the secrecy performance of future wireless networks (5G and beyond). Many contributions have been made in this domain; here we summarize some of the key works and studies in order to (1) highlight the potential of this new emerging area of research for boosting and improving the delivered secrecy and (2) discover and figure out the potential future research topics that can be pursued further [38].

In [39], a detailed secrecy performance analysis was performed on both noise-limited and artificial noise (AN) assisted mm-Wave networks using the stochastic geometry framework. For both cases, the non-colluding and colluding eavesdroppers' scenarios were assumed. In [40], the characterization of the secrecy outage considering blockages at Bob and Eve was evaluated in a mm-Wave network coexisting with a microwave network. In [41] and [42], authors explained how to apply the AN (artificial noise)-based scheme to mm-Wave systems and then analyzed the obtained secrecy performance. In [43], three secure transmission schemes were comprehensively investigated over an mm-Wave channel by evaluating both the secrecy outage probability and the secrecy throughput. Particularly, the authors analyzed the new secrecy properties of the maximum ratio transmission (MRT) for the mm-Wave system. Then, another two secure transmission schemes were introduced, namely AN-based beamforming and partial MRT-based (PMRT) beamforming.

In [44], a simple directional modulation technique, called antenna subset modulation (ASM), was introduced by utilizing the potential of massive antenna arrays to provide secrecy. It was revealed that ASM achieves coherent symbol detection at the desired direction, while ensuring a high error rate in the undesired directions. In [45], the secrecy throughput was analyzed from the perspectives of delay-tolerant and delay-limited transmission schemes, using analog beamforming with phase shifters to reduce the system cost. The authors of [46] proposed a new wireless transmission technique, called silent antenna hopping (SAH) or low-complexity ASM [47], to further enhance the achievable secrecy. SAH is composed of a phased-array transmitter followed by antennas with an on-off switching circuit. This scheme ensures scrambling the constellation points in both amplitude and phase in the undesired signal direction.

In [48], the authors generalized the scheme proposed in [46] and [47] to a new architecture, called switched phased-array (SPA), to further enhance the physical security. SPA was shown to work as a platform for three different transmission techniques: (1) conventional phased-array transmission; (2) antenna subset transmission (AST) technique; and (3) silent antenna hopping (SAH) transmission technique. In [49], the authors proposed a transmission scheme which can provide more security than the ASM scheme proposed in [47] and with low computational complexity. This was attained by designing an optimized antenna subset selection based on an iterative fast Fourier transform (FFT). The physical layer security performance in mm-Wave ad-hoc networks was explored in [50], where eavesdroppers are randomly located, and can intercept confidential messages as they may be situated in the main signal beam. Particularly, the authors characterized the impact of mm-Wave channel features, random blockages, and antenna gains on the secrecy performance.

The authors of [51] thoroughly studied physical layer security in a MISO mm-Wave scenario in which multiple eavesdroppers with a single antenna are randomly situated in the network. In this study, both maximum ratio transmitting (MRT) beamforming and artificial noise (AN) beamforming are investigated. It is reported that the achievable secrecy level is closely affected by eavesdroppers' density in the network as well as the spatially resolvable paths of both Bob's and Eve's channels in mm-Wave systems. The work in [52] studied secure beamforming (BF) for fifth-generation (5G) cellular systems operating at the mm-Wave frequency and coexisting with a satellite network. In this work, a uniform planar array is employed at the base-station where it is assumed that the channel state information of multiple eavesdroppers based on imperfect angle-of arrival (AoA) is known. Under the constraint of the interference and transmit power of the base station (BS), authors formulated a constrained optimization problem in order to maximize the secrecy rate of worst case of the cellular user. Afterward, they also proposed two beamforming methods to solve the optimization problems for the case of uncoordinated and coordinated Eves. In [53], the authors proposed an interesting PLS scheme for mm-Wave communication based on a hybrid MIMO phased-array time modulated DM. The basic idea is to divide the transmit array into multiple sub-arrays. All the sub-arrays can jointly work as a MIMO for multi-user communications or higher angular resolution while each sub-array forms a secure directional beam. More importantly, effective security can be achieved by applying a time-modulated DM scheme for phased MIMO mm-Wave wireless communication without having knowledge of eavesdroppers.

4.5 Open Research Issues

Despite the varying advantages offered by the mm-Waves, they suffer from a number of critical problems that must be resolved. There is a need to distinguish between different standards for broader market exploitation; the IEEE 802.15.3c is positioned to provide gigabit rates and longer operating range [12]. At these rates and ranges, it is a nontrivial task for millimeter -wave systems to provide sufficient power margin that assures a reliable communication link. Furthermore, delay spread across the communication channel is another limiting factor for high-speed transmissions.

Meanwhile, the emergence of mm-Wave communications has certainly created the need for new signal processing, channel modeling, circuit, antenna, and communication designs and security aspects. As seen from the above literature, many of the proposed approaches for securing mm-Wave systems do not pay much attention to the fact that applying security techniques in the digital baseband domain (like conventional systems) is extremely challenging in mm-Waves and should be avoided as much as possible due to issues related to power, complexity, and cost of future IoNT devices. Therefore, there is a need for devising new practical security designs that can be implemented in the analog domain instead of using purely digital domain-based schemes. Recently, there have been some emerging studies on hybrid analog-digital designs for physical layer security in mm-Waves as can be found in [52–55].

4.6 Conclusion

mm-Waves can introduce low-interference millimeter wave communications as a promising technology to alleviate the pressure of scarce spectrum resources for next-generation wearable devices in the IoNT paradigm. Several types of the front-end smart antennas for mm-Wave communication systems have been overviewed. Design aspects of the mm-Waves' antennas have been investigated in heterogeneous large-scale multiple-antenna deployments. It supports the future development of next-generation networks by employing the large-scale multiple antennas technology in mm-Wave communications. We would like to remark also that the considered mm-Wave technology in wearable devices can vary from those listed so far in the literature tackling the monitoring problem from the wireless body area network (WBAN) perspective. Where we believe significant enhancement to the IoNT-based paradigm in the industry and healthcare sectors can be achieved is via integration with the abundance of smart devices/sensors around us. This vision can employ intelligent and cognitive approaches with the help of cloud/edge paradigms in delivering multimedia data that satisfies the end-user requirements and service providers' expectations/budget.

References

1. Zhifeng H., Mao S., Rappaport T.S. 2015. On link scheduling under blockage and interference in 60-GHz ad hoc networks access." *IEEE*, 3, 1437–1449.
2. Moon Y.J., Kim W.W. 2015. SNR weighted LLR combining method in uplink mmWave environment. IEEE International Conference on Information and Communication Technology Convergence (ICTC), Jeju, South Korea.
3. Deng S., MacCartney Jr. G.R., and Rappaport T.S. 2015. Indoor office plan environment and layout-based mmWave path loss models for 28 GHz and 73 GHz. arXiv preprint arXiv:1511.07057.
4. Boccardi F., Heath R.W., Lozano A., Marzetta T.L., Popovski, P. 2014. Five disruptive technology directions for 5G. *Communications Magazine, IEEE*, 52(2), 74–80.
5. Andrews, J.G., Buzzi, S., Choi W., Hanly S.V., Lozano A., Soong A.C., Zhang J.C. 2014. What will 5G be? Selected areas in communications, *IEEE Journal on Selected Areas in Communications*, 32(6), 1065–1082.
6. Rangan S., Rappaport T.S., Erkip E. 2014. Millimeter-wave cellular wireless networks: Potentials and challenges. *Proceedings of the IEEE*, 102(3), 366–385.
7. Wonil R., Seol J.Y., Park J., Lee B., Lee J., Kim Y., Cho J. 2014. Millimeter-wave beamforming as an enabling technology for 5G cellular communications: Theoretical feasibility and prototype results. *Communications Magazine, IEEE*, 52(2), 106–113.
8. Bhushan N., Li J., Malladi D., Gilmore R., Brenner D., Aleksandar Damnjanovic A., Sukhavasi R., Patel C., Geirhofer S. 2014. Network densification: The dominant theme for wireless evolution into 5G. *Communications Magazine, IEEE*, 52(2), 82–89.
9. Kukutsu N., Kado Y., March 2009. Overview of millimeter and terahertz wave application research. *NTT Technical Review*, 7(3), 1–6.
10. Al-Turjman F., Imran M., Vasilakos A. 2017. Value-based caching in information-centric wireless body area networks, *Sensors Journal*, 17(1), 1–19.
11. Al-Turjman F. 2017. A rational data delivery framework for disaster-inspired internet of nano-things (IoNT) in practice, *Springer Cluster Computing*. DOI: 10.1007/s10586-017-1357-7.
12. Oliver A.D. 1989. Millimeter wave systems—past, present and future. *IEEE Proceedings*, 136(1), 35–52.
13. IEEE802.15 WPANTM Terahertz Interest Group. http://www.ieee802.org/15/pub/1Gthz.html.
14. IEEE 802.15 WPAN Millimeter Wave Alternative PHY Task Group 3c (TG3c).
15. Fisher R. 2007. 60 Ghz WPAN standardization within IEEE 802.15.3c. 2007 International Symposium on Signals, Systems and Electronics, Montreal, QC, Canada.
16. Hirata A., Yamaguchi R., Sato Y., Mochida T., Shimizu K. 2006. Multiplexed transmission of uncompressed HDTV signals using 120-GHz-band millimeter-wave wireless communications system. *NTT Technical Review*, 4(3), 64–70.
17. Rappaport T.S. 2002. *Wireless communications: Principles and practice*, 2nd ed. Englewood Cliffs, NJ: Prentice Hall.
18. Heath R.W., González-Prelcic N., Rangan S., Roh W., Sayeed A.M. April 2016. An overview of signal processing techniques for millimeter wave MIMO systems. *IEEE Journal of Selected Topics in Signal Processing*, 10(3), 436–445.
19. Williamson M.R., Athanasiadou G.E., Nix A.R. 1997. Investigating the effects of antenna directivity on wireless indoor communication at 60 GHz. in *Proc. IEEE Int. Symp. Pers. Ind. Mobile Radio Commun. (PIMRC)*, 2, 635–639.

20. Tolbert C., Straiton A. 1966. Attenuation and fluctuation of millimeter radio waves. *Proc. IRE Int. Conv. Rec.*, 5, 12–18.

21. Giannetti F., Luise M., Reggiannini R. 1999. Mobile and personal communications in 60 GHz band: A survey. *Wirelesss Pers. Comun.*, 10, 207–243.

22. Xu H., Kukshya V., Rappaport T.S. April 2002. Spatial and temporal characteristics of 60 GHz indoor channel. *IEEE J. Sel. Areas Commun.*, 20(3), pp. 620–630.

23. Zwick T., Beukema T.J., Nam H. July 2005. Wideband channel sounder with measurements and model for the 60 GHz indoor radio channel. *IEEE Trans. Veh. Technol.*, 54(4), 1266–1277.

24. Anderson C.R., Rappaport T.S. May 2004. In-building wideband partition loss measurements at 2.5 and 60 GHz. *IEEE Trans. Wireless Commun.*, 3(3), 922–928.

25. Lu J.S., Steinbach D., Cabrol P., Pietraski P. December 2012. Modeling human blockers in millimeter wave radio links. *ZTE Commun.*, 10(4), 23–28.

26. Schwering F.K., Violette E.J., Espeland R.H. May 1988. Millimeter-wave propagation invegetation: Experiments and theory. *IEEE Trans. Geosci. Remote Sens.*, 26(3), 355–367.

27. Ben-Dor E., Rappaport T.S., Qiao Y., Lauffenburger S.J. 2001. Millimeter-wave 60 GHz outdoor and vehicle AOA propagation measurements using a broadband channel sounder. In *Proc. IEEE Global Telecommun. Conf. (GLOBECOM)*, 1–6.

28. Rappaport T.S., Sun S., Mayzus R., Zhao H., Azar Y., Wang K., Wong G.N., Schulz J.K., Samimi M., Gutierrez F. May 2013. Millimeter wave mobile communications for 5G cellular: It will work! *IEEE Access*, 1, 335–349.

29. Bai T., Vaze R., Heath Jr., R.W. September 2014. Analysis of blockage effects on urban cellular networks. *IEEE Trans. Wireless Commun.*, 13(9), 5070–5083.

30. Kulkarni M., Singh S., Andrews J. December 2014. Coverage and rate trends in dense urban mmWave cellular networks. *Proc. IEEE Global Telecommun. Conf. (GLOBECOM)*, 3809–3814.

31. Bai T., Heath Jr., R.W. February 2015. Coverage and rate analysis for millimeter wave cellular networks. *IEEE Trans. Wireless Commun.*, 14(2), 1100–1114.

32. Tse D., Viswanath P. 2007. *Fundamentals of wireless communication*. Cambridge, UK: Cambridge University Press.

33. Kim C., Kim T., Son J.-S., Seol J.-Y. April 2014. On the hybrid beamforming with shared array antenna for mmWave MIMO-OFDM systems. *Proc. IEEE Wireless Commun. Netw. Conf. (WCNC)*, 335–340.

34. Alkhateeb A., Jianhua M., González-Prelcic N., Heath Jr., R.W. December 2014. MIMO precoding and combining solutions for millimeter-wave systems. *IEEE Commun. Mag.*, 52(12), 122–131.

35. Brady J., Behdad N., Sayeed A.M. July 2013. Beamspace MIMO for millimeter-wave communications: System architecture, modeling, analysis and measurements. *IEEE Trans. Antennas Propag.*, 61(7), 3814–3827.

36. Sayeed A.M., Behdad N. September 2010. Continuous aperture phased MIMO: Basic theory and applications. *Proc. Annu. Allerton Conf. Commun. Control Comput.*, 1196–1203.

37. Güvenkaya E., Hamamreh J.M., and Arslan H. December 2017. On physical-layer 3715 concepts and metrics in secure signal transmission. *Phys. Commun.*, 3716(25), 14–25.

38. Hamamreh J.M., Furqan H.M., Arslan H. 2018. Classifications and applications of physical layer security techniques for confidentiality: A comprehensive survey. IEEE Communications Surveys and Tutorials.

39. Wang C., Wang H.M. August 2016. Physical layer security in millimeter wave cellular networks. *IEEE Trans. Wireless Commun.*, 15(8), 5569–5585.
40. Vuppala S., Biswas S., Ratnarajah T. August 2016. An analysis on secure communication in millimeter/micro-wave hybrid networks. *IEEE Trans. Commun.*, 64(8), 3507–3519.
41. Ju Y., Wang H.M., Zheng T.X., Zhang Y., Yang Q., Yin Q. September 2016. Secrecy throughput maximization for millimeter wave systems with artificial noise. *2016 IEEE 27th Annual Int. Symp. on Personal, Indoor, and Mobile Radio Commun. (PIMRC)*, 1–6.
42. Ju Y., Wang H.M., Zheng T.X., Yin Q. April 2016. Secure transmission with artificial noise in millimeter wave systems. *2016 IEEE Wireless Commun. and Netw. Conf.*, 1–6.
43. Ju Y., Wang H.M., Zheng T.X., Yin Q. May 2017. Secure transmissions in millimeter wave systems. *IEEE Trans. Commun.*, 65(5), 2114–2127.
44. Valliappan N., Lozano A., Heath R.W. 2013. Antenna subset modulation for secure millimeter-wave wireless communication. *IEEE Trans. Commun.*, 61(8), 3231–3245.
45. Wang L., Elkashlan M., Duong T.Q., Heath R.W. June 2014. Secure communication in cellular networks: The benefits of millimeter wave mobile broadband. *2014 IEEE 15th Int. Work. Sig. Process. Advances Wireless Commun. (SPAWC)*, 115–119.
46. Alotaibi N.N., Hamdi K.A. December 2015. Silent antenna hopping transmission technique for secure millimeter-wave wireless communication. *2015 IEEE Glob. Commun. Conf. (GLOBECOM)*, 1–6.
47. Alotaibi N.N., Hamdi K.A. April 2016. A low-complexity antenna subset modulation for secure millimeter-wave communication. *2016 IEEE Wireless Commun. Netw. Conf.*, 1–6.
48. Alotaibi N.N., Hamdi K.A. March 2016. Switched phased-array transmission architecture for secure millimeter-wave wireless communication. *IEEE Trans. Commun.* 64(3), 1303–1312.
49. Chen C., Dong Y., Cheng X., Yi N., January 2017. An iterative FFT-based antenna subset modulation for secure millimeter wave communications. *2017 Int. Con. on Comp., Netw. Commun. (ICNC)*, 454–459.
50. Zhu Y., Wang L., Wong K.K., Heath R.W. May 2017. Secure communications in millimeter wave ad hoc networks. *IEEE Trans. Wireless Commun.*, 16(5), 3205–3217.
51. Ramadan Y.R., Minn H. November 2017. Artificial noise aided hybrid precoding design for secure mm wave MISO systems with partial channel knowledge. *IEEE Sig. Proc. Lett.*, 24(11), 1729–1733.
52. Ramadan Y.R., Minn H., Ibrahim A.S. November 2017. Hybrid analog-digital precoding design for secrecy mm-wave MISO-OFDM systems. *IEEE Trans. Commun.*, 65(11), 5009–5026.
53. Tian X., Li M., Wang Z., Liu Q. December 2017. Hybrid precoder and combiner design for secure transmission in mmwave MIMO systems. *GLOBECOM 2017– IEEE Global Commun. Conf.*, 1–6.
54. Ju Y., Wang H.M., Zheng T.X., Yin Q., Lee M.H. 2018. Safeguarding millimeter wave communications against randomly located eavesdroppers. *IEEE Trans. Wirel. Commun.*, 17(4), 2675–2689.
55. Vuppala S., Tolossa Y.J., Kaddoum G., Abreu G., March 2018. On the physical layer security analysis of hybrid millimeter wave networks. *IEEE Trans. Commun.*, 66(3), 1139–1152.

Chapter 5

A Rational Routing Protocol for WBAN

Fadi Al-Turjman

Department of Computer Engineering, Antalya Bilim University, Antalya, Turkey

Contents

5.1 Introduction 60
5.2 Related Work 61
5.3 System Models 63
    5.3.1 Network Architecture 63
    5.3.2 Lifetime in IoNT 64
    5.3.3 Energy Conservation and Dead Node Issue....................................65
    5.3.4 Communication Model 65
    5.3.5 Traffic Model 66
5.4 Rational Data Delivery Framework..67
    5.4.1 Learning 68
    5.4.2 Reasoning 70
5.5 Performance Evaluation 73
    5.5.1 Shortest Path Algorithm (SPA) 73
    5.5.2 Nearest Neighbor Algorithm (NNA)..74
    5.5.3 Experimental Setup 75
    5.5.4 Threats to Validity 76
    5.5.5 Performance Metrics and Parameters..76
    5.5.6 Simulation Results 77
5.6 Conclusions...83
References 83

5.1 Introduction

In risk management, nanotech has played a significant role in detection and containment of disasters [1]. In chemical engineering, for instance, carbon nanotubes have been used to sniff out dangerous and toxic gases; a network of these sensors can be laid out and used to monitor the motion of toxic gases over a large area [2]. In medicine, the existence of a disinfectant that works better and more efficiently than conventional traditional ones, by providing long lasting anti-viral effect against major viruses, has proven the importance of nanosensing technology in disaster management [3]. The above-mentioned examples are just few of the many areas where nanosensing technology has made massive improvements. However, this technology is still suffering extreme limitations in terms of connectivity while collaborating in wireless network-based systems.

The connectivity and links between nanodevices (nodes) distributed to monitor a specific phenomenon have led to the idea of nanonetworks followed by the Internet of Nano-Things (IoNT) proposal. In IoNT, nanonetworks are connected to the internet via local gateways. Integrating IoNT with other local area network systems will expand the array of services that can be provided to public users as well as decision makers.

IoNT stands out in terms of its distinctive features related to limited-energy constraints, short communication range in the THz band, and low processing power, and needs to be assimilated into the routing protocols so as to realize this new paradigm. Different challenges that face the data-routing process in IoNT are still being looked into, but a complete, effective solution has not been developed yet. Nanonetworks consume energy on all levels of their processes; they consume energy while sensing data, transmitting data, and processing data. Wireless multi-hop networks were used to achieve energy efficiency in such networks; consequently, adequate schemes have been proposed [4–7]. Nevertheless, such schemes end up being useless, unable to be used in real-life scenarios because they assume a static network topology [10–12], whereas nanonetworks show a haphazard network topology due to the mobility of the nanosensors' carrier or because they are restricted to two-hop routing schemes.

Design and implementation of routing algorithms are considered imperative in nanonetworks. This is because nanonetworks' sensors are usually restricted in their processing power, communication range, and energy aptitudes. In this chapter, we propose a rational data delivery approach (RDDA) that addresses the challenges encountered in the IoNT paradigm. Two major features have been considered while realizing rationality (cognition) in this approach; these are reasoning and learning. Reasoning is used to determine and prioritize the characteristics of a given traffic flow and select the next hop for data transmission along the direction of the data flow. Whereas reasoning helps in realizing short term objectives and helping the network improve its current status, learning is used to accomplish long-term goals such as improving the lifetime of the network. The response obtained from the history of the network helps in the learning process and also helps in planning preemptive feedback. Hence, the proposed RDDA approach is energy efficient and is designed to enhance the current status of the nanonetwork, and thus, assure quality of information (QoI). Additionally, in order to deal with the dynamicity of the nanonetwork topology, the proposed approach is verified via a grid-based distribution of the nanosensors on a monitored mobile object/body for effective observations and conclusions [15].

The remainder of this chapter is organized as follows. Section 5.2 reviews previous related studies. Section 5.3 discusses our system models. Section 5.4 describes our proposed routing approach for IoNT paradigm. Section 5.5 provides performance evaluation for the proposed approach. Finally, Section 5.6 provides the conclusions and future directions.

5.2 Related Work

The communication in nanonetworks can utilize one of the following technologies; nanomechanical, acoustic, electromagnetic, and chemical or molecular communication [8]. Mainly due to their tiny sizes, nanonetworks introduce difficulties in both hardware and software design. Especially for the software part, the communication layer stack needs fine tuning as such tiny hardware imposes critical restrictions. Knowing that the physical wireless signaling is performed at THz bands, due to the restricted antenna size, this necessitates special routing/communication techniques [9].

Routing protocols in nanonetworks can be classified into simple flooding and random point-to-point protocols. These protocols can be optimized and customized for more efficient performance. However, several design aspects shall be taken in to consideration, such as the nanonetwork topology, nodes' mobility, deployment space (2D vs. 3D), and energy. In fact, energy is the most significant and limiting factor according to current nanotechnology studies [2]. In that sense, routing protocols, which optimize the energy consumption in nanonetworks while satisfying different constraints, are expected to have a great influence on the IoNT paradigm. Existing routing approaches in nanonetworks aim at extending the network lifetime by minimizing the energy consumption while considering traditional metrics which might not be effective in practice. In [16], authors proposed a peer-to-peer routing protocol. In their work, 2D uniform grids and 2D uniform random topologies are assumed, in which identical nanosensors are deployed. Packet collisions and redundant retransmissions are the only two metrics that have been considered while optimizing the proposed protocol. In this protocol, nodes are classified based on the packet reception statistics they have logged. The routing scheme exploits this classification in optimizing energy consumption. In [17], coordinate-based addressing scheme is proposed for nanosensors distributed uniformly in a rectangular 2D topology. The proposed routing protocol tries to minimize the hop count of the packet transmission by placing anchor nodes at the vertices of the grid. This routing protocol is assessed by considering packet retransmission rate, successful packet reception rate, and packet loss rate. In [18], channel-aware routing protocol is proposed. Authors considered the special attributes of the THz band communication. The forwarding is optimized by considering two cost factors: namely, avoiding long-distance region in which the signal may suffer the path loss and avoiding short-distance region in which the number of hops can be increased dramatically. However, their achieved results are based on simple 1D simulations. Authors in [19] focused on the physical layer part for their routing protocol. They proposed a physical network coding routing protocol by extending a geographical greedy routing algorithm for nanonetworks. The packets are separated into two parts and transmitted in pairs along pipelined multi-hop route, while avoiding grouped weak nodes to achieve energy effectiveness. The work presented in [20] proposes a geographic routing protocol; nodes of the nanonetworks are assumed to comprise two types of anchors, which have higher communication and processing capabilities than the edge nodes. Edge nodes are required to localize their positions in order to reference these anchor nodes. The authors assume that the network topology is square with four anchors located at the corners. Their routing approach operates in two phases: the setup phase and the operation phase. The setup phase is designed to assist the network edge nodes in measuring their distances from the anchors. In the operation phase, a source node selects anchor nodes and incorporates this information in transmitted packets' headers. A receiving node checks its location, the destination location, and the source location to decide on forwarding or dropping the packet. However, this approach requires addressing for all nodes, which forms a significant challenge in the IoNT with nanoscale applications. The work in [22] presents a flooding data dissemination scheme. The proposed scheme assumes a square grid network architecture where the nanosensors are distributed densely at the vertices of the grid. Utilizing the uniform nodes' patterns and lattice algebra, the scheme dismisses the requirement for node addressing and employs a simple flooding routing scheme for data dissemination. The scheme relies on classifying each node as either an infrastructure or single user node, depending on its reception quality.

Based on the previously discussed attempts in the literature, we can conclude that energy-aware routing protocols aim at identifying the shortest path and/or the nearest neighbor towards destination in the nanonetwork. In the nearest neighbor approach (NNA), when a packet is transmitted from one node to another, it follows the shortest path [13]. NNA assumes that if a packet always follows shorter path, it will use shortest path until it reaches destination node. In short, this algorithm uses four-direction transmission (left, right, up, down) only in virtual grid setups, where the closest vertical/horizontal but not diagonal relying neighbor is used to send the data packet [14]. As a result, the hop count can unnecessarily increase and also the energy consumption is negatively affected by increased hop count. Meanwhile, in the shortest path approach (SPA), when a data packet is transmitted from a node, it calculates the shortest path from the sender node to the destination instead of the node-to-node fashion. Accordingly, SPA uses eight-direction data forwarding (up, upper-left, upper-right, down, down-left, down-right, right and left), and thus, it considers the shortest path to destination rather than shortest neighbor to relay. Nevertheless, nanosensors in the targeted IoNT can typically follow a random behavior. They can move around the human body for certain health applications, and therefore, may need to be associated with varying neighbors frequently, and hence, may not always have a fixed network structure/topology. In this research, we proposed a rational data delivery algorithm (RDDA) as a distinguished routing protocol for the IoNT. It assumes a multitier nanonetwork and cluster/tier-wide synchronization. Moreover, it's a topology-independent protocol which copes with the randomness nature in nanonetworks. According to RDDA, the system determines the path from the routing node (RN) to the destination node in view of each node's remaining energy. The remaining energy of recent RN's neighbors is controlled each time before a data packet is sent from the RN. If one of these neighboring RNs' energy is below half of the initial energy, a new alternative path will be determined and the data packet will be forwarded accordingly. Although this can increase the hop count in comparison to SPA, the energy efficiency will be improved and network lifetime will be prolonged.

5.3 System Models

IoNT in smart environments emerges to control physical/chemical changes and pass the information to sophisticated data centers for processing [23]. In smart environments, many parameters, such as pressure, temperature, sound, etc. can be sensed by the IoNT paradigm. Each nanosensor in the IoNT has different capabilities such as sensing, processing, and communicating. Due to energy constraints, these nanosensors cannot communicate with each other, and the only way in which they serve is to pass the sensed data to a nanorouter (NR). NRs send the collected data to a gateway (GW) that is usually connected to the Internet for remote control. Although nanonetworks have several advantages, there are also many challenges that need to be considered carefully. One of the most important challenges is energy consumption. Therefore, an energy-efficient routing protocol is a key factor in prolonging the utilized nanonetwork lifetime dramatically. In the following section, we describe the assumed system model for the proposed RDDA approach.

5.3.1 Network Architecture

With the networking technology, nanosensors have more potential, since they can cooperate and communicate to achieve more challenging tasks. Figure 5.1 shows the general network architecture to be assumed in this chapter for the vision of

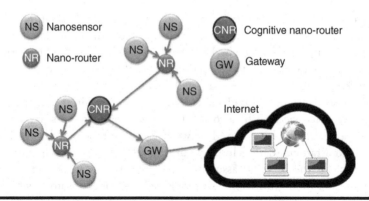

Figure 5.1 Network architecture and main components in the IoNT.

the IoNT paradigm. Significant elements of the nanonetworks are the nanosensors, NRs, and cognitive nanorouters (CNRs). Nanosensors are the smallest and simplest nanodevices. These devices can only perform simple computation tasks and can transmit over very short distances due to limited energy and memory and reduced communication capabilities. NRs have slightly larger computational resources than nanosensors, and thus can aggregate information. CNRs, also called nano-micro interfaces, are used to further aggregate the information forwarded by the NR and send them to a micro-scale device. And thus, CNRs are hybrid devices which can communicate in the nanoscale and can utilize classical communication paradigms in micro- and/or macrocommunication networks. Through GWs these types of networks can be connected to the traditional Internet.

The communication range in IoNT is predicted to be between 1 nm and 1 cm in terahertz-band [24]. And thus, multi-hop routing is an effective data delivery style. Moreover, the direction of a communication route is not deterministic and depends on the drift velocity of nanosensors, which may result in service disruption and extended delays [25].

5.3.2 Lifetime in IoNT

Lifetime in this research is defined as the time or number of transmission rounds in which the nanonetwork can no longer send useful information to the end users. It is reflected by the network's inability to find a path to deliver data with satisfactory values for a number of QoI attributes such as latency, fairness, and remaining energy [21]. This definition provides information in order to satisfy the application requirements and considers the status of the network and nanoresources in defining the lifetime of the network. Moreover, it gives an acceptable explanation for the fact that if the network does not have required resources to deliver packets, it cannot satisfy the end user and thus, it should be considered as a dead nanonetwork. Therefore, we can evaluate the lifetime of the nanonetwork in the IoNT by either counting the alive nanosensors [26], checking the ratio of still-covered areas to the uncovered ones by the nanonetwork, or based on both [27].

5.3.3 Energy Conservation and Dead Node Issue

Energy in nanonetworks can be a critical factor towards realizing the main objective of the emerged IoNT paradigm. Knowing that majority of the nanonetwork energy budget is spent on routing data, we focus in this study on the NR energy expenditures. According to [24] this can be characterized by the following equation.

$$E_{NR} = C(T * (E_{TX}) + R * (E_{RX})) \tag{5.1}$$

where E_{TX} and E_{RX} are transmission and reception energy, respectively. C indicates the cost function of the energy consumed, and T and R are the number of transmitted and received packets, respectively. As discussed earlier, the main function of CNR is data aggregation and routing of traffic received from the NRs. Therefore, it is expected that CNRs consume additional energy compared to regular NRs. This energy consumption can be characterized as follows:

$$E_{CNR} = C(T * (E_{TX}) + R * (E_{RX})) + C(A * (E_{ag})) + C(P * (E_{cog} - E_{pro})) \tag{5.2}$$

In Eq. (5.2), A and P, represents the total number of packets that are aggregated and processed by the cognitive nanorouters, respectively. $C(A * (E_{ag}))$ shows the energy cost during data aggregation, and $C(P * (E_{cog} - E_{pro}))$ reveals the energy cost due to protocol and processing overhead while performing cognitive (rational) processes. By forming Eq. (5.2) in terms of the energy cost of NRs we obtain:

$$E_{CNR} \geq E_{NR} + C(A * (E_{ag}) + C(E_{cog} - E_{pro}) \tag{5.3}$$

If the NR and CNRs use the same transmit power, the equality sign becomes positive in Eq. (5.3). In this study, we assume multi-tier NRs' distribution. Once all of the first tier NRs are dead, no other node will be able to send data to the GW, and the lifetime of the network will be over. Figure 5.2 shows the circular distribution of dead and alive nodes in such a situation.

5.3.4 Communication Model

It is of great importance to study the communication nature in very short range because in this case it is functioning at the nanoscale [28, 29]. Therefore, the proposed path loss formula in [28] at THz is considered. This formula has two essential parts: (1) the absorption path loss, and (2) the spread path loss. In general, energy-aware frameworks are heavily dependent on two main principles in their communication design: (1) the number of hops without delay constraints, and (2) the

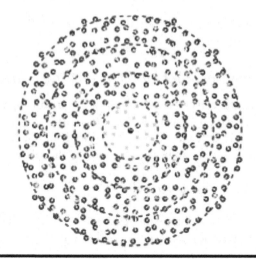

Figure 5.2 Many nodes are alive; but, as soon as the last two surviving nodes in the first-tier die, the network will be disconnected [24].

number of hops with delay constraints. If the communication system involves multiple hops, processing, buffering and transmitting/receiving delays are repeated a number of times equal to the hop count [21].

5.3.5 Traffic Model

There are three types of data traffic that a nanonetwork can experience in the assumed IoNT paradigm. These types are on-demand, periodic, and emergency traffic. Each of them is associated with a different QoI requirement based on the served IoNT application. The QoI requirement for the routed data can be decided by the following attributes: (1) the network reliability (fairness), (2) the end-to-end delay/latency, and (3) the energy consumed. As shown in Table 5.1, we are not allocating absolute numbers on the attribute; rather, we associate priorities with each attribute and let the priorities decide on the importance of the attribute's absolute value. Values in Table 5.1 show the associated priority with the attribute. Number 1 indicates the highest priority while number 3 indicates the lowest.

Table 5.1 Priority per QoI Attribute in IoNT

Attribute	Priority
Lifetime (or Energy)	1
Delay	2
Fairness (or reliability)	3

Assuming Markov queuing model and the Poisson process packet arrival rate λ, all transitions from an empty-queue status to a non-empty status accordingly can be represented by:

$$P_{0,i} = \lambda_i * \psi, \forall i = 0,\dots,L \tag{5.4}$$

where $P_{i,j}$ is the probability of the node status change from state i to state j, ψ is the inter-arrival time of a new packet to the node, and L is the queue/buffer length at the IoNT node. Accordingly, the probability of having a non-decreasing queue (i.e., non-transition probability) can be achieved by:

$$P_{i,i-1} = \beta * \psi * \lambda_{j-i+1} * \psi + (1 - \beta * \psi) * \lambda_{j-i} * \psi, \quad \forall i = 1,\dots,L-1 \tag{5.5}$$

$$P_{i,L} = \beta * \psi * \lambda_{\geq L-i+1} * \psi + (1 - \beta * \psi) * \lambda_{\geq L-i} * \psi, \quad \forall i = 1,\dots,L \tag{5.6}$$

where β is the probability of transmitting a data packet from a nanosensor in the nanonetwork. Now, let us assume the Markov model with the aforementioned finite set of transitions and transitional probabilities $\pi_w(t)$ in matrix P of having m packets. The steady state equations for each data traffic packet can be described as follows:

$$\pi_0(t) = \pi_0 * (1 - \lambda * \psi) + \pi_1 * \beta * \psi + o(\psi) \tag{5.7}$$

$$\pi_L(t) = \pi_{L-1}(t) * \beta * \psi + \pi_L(t) * (1 - \beta * \psi) + o(\psi) \tag{5.8}$$

$$\begin{aligned} \pi_w(t) = \pi_{w-1}(t) * \beta * \psi + \pi_w(t) * (1 - \lambda * \psi - \beta * \psi) \\ + \pi_{w+1}(t) * \beta * \psi + o(\psi), \quad \forall w! = 0 \end{aligned} \tag{5.9}$$

where $\pi_i(t)$ denotes the steady state of the packet i at time t, and $o(\psi)$ is defined as a function of ψ such that $\lim_{\psi \to 0} \frac{o(\psi)}{\psi} = 1$. Since the proposed model is considered to be an irreducible, periodic, and recurrent non-null Markov chain, the model holds the unique stationary probability $\Pi_w(t) = \{\Pi_0(t),\dots, \Pi_L(t)\}$ where $\Sigma_{w=0}^L \Pi_w(t) = 1$, which strictly provides that the mean rate of arrivals per state is less than the mean rate at which packets are obtained by the IoNT node per state.

5.4 Rational Data Delivery Framework

In this section we propose a novel rational data delivery approach (RDDA) for the IoNT paradigm. Assume x is a randomly selected nanosensor by the GW based on the required data in a specific IoNT application. The random number of relays within the communication range of the nanosensor x can be modeled by a spatial Poisson process X[30]. Assume that the nanosensor x can be at point $z \in \mathbb{R}^2$ and

$l(z,X)$ is the shortest distance from z to the nearest point of X such that $l(z,X) \leq r$. Since X is a spatial Poisson process, then $l(z,X) \leq r$, if and only if $NR(d(z,r)) > 0$, where $d(z,r)$ is a disc of radius r centered at z. And $NR(d(z,r))$ is a Poisson random variable denoting the number of nanorouters within the disk $d(z,r)$ with remaining energy sufficient to transmit at least once. Consequently, the probability of having at least one NR neighbor within the transmission range of the nanosensor x is given as follows.

$$P(l(z,X) \leq r) = P(NR(d(z,r)) > 0)$$
$$= 1 - P(NR(d(z,r)) = 0) \quad (5.10)$$

In this study, it is assumed that the nanonetwork is dead when the lifetime of the neighboring NRs is expired. Thus, assuming $f(x_j)$ is the cost function of transmitting from NR_j to GW in terms of fairness, $g(x)$ is the energy of neighboring NRs, $h(x)$ is the minimum distance from a neighbor NR_j to GW, $i(x)$ initial energy of the neighboring NR. Accordingly, the RDDA framework assumes three main criteria for data routing; (1) evaluation criteria; $f(x_j) = $ Cost(Neighbor NR to GW) and $h(x_j) = min(f(x_j))$, this is guaranteed by lines 11 to 18 in Algorithm 5.1, (2) selection criteria; $g(h(x_j)) > i(h(x_j))*50\%$, is found between lines 19 and 21, and (3) termination criteria; all one-hop NRs are dead or $P(l(z,X) \leq r) = 0$.

In Algorithm 5.1, rational (cognitive) elements such as reasoning and learning are applied at the CNR. In the following a detailed description about these elements is provided.

5.4.1 *Learning*

Learning is used in our RDDA approach in order to identify the most appropriate routes toward the Internet gateway while maintaining several QoI attributes in the nanonetwork, such as fairness, delay, and energy-efficiency. Via learning, each time a CNR has to choose an NR on the route, it excludes NRs which can increase the cost in terms of QoI attributes between the current NR and the gateway. Positions of those NRs which best fit the required QoI in a nanonetwork are saved in the CNR for future use as well. Thus, the direction, along with the destination feedback about the chosen path, helps the CNRs to learn and improve paths toward destinations in the IoNT. In the following, we elaborate more on this cognitive feature/element through an illustrative example.

Example 1. Let's assume we have n nanorouter, where the ith available router $R_i \in \{R_1, R_2,..., R_n\}$. S_1 and S_2 have a data packet to be sent to destination devices D_1 and D_2. Out of these relays, it is determined that R_5 provides the lowest cost to D_1 and D_2 as shown in Figure 5.3(a). Therefore, S_1 sends the data packet to R_5. And S_2 sends its data packets to R_5, as well. As a result, the route through R_5 becomes

Algorithm 5.1   The RDDA Approach

---

1 Function RDDA

2 Input

3   A set of NRs connected to the nanosensor (sender).

4 Output

5   NR_j chosen by RDDA to deliver data towards GW.

6 Begin

7 Initialize

8   Hop Count = 0;

9   **For** each $NR_i \in$ {NRs}

10   Identify NR_i as a start node for current round.

10   List all neighboring NR from source NR_i

11   **If** NR_i has one-hop to GW, send directly.

12   **Else If** there is least one NR connected with NR_i

13   **For** each source NR index 'j' do

14                $f(NR_j) =$ Distance

15                $g(NR_j) =$ Energy

16                $h(NR_j) = min(f(NR_j))$

17                $i(NR_j) =$ Initial Energy

18   **End**

19   **If** $g(h(NR_j)) > i(h(NR_j)) * 50\%$

20   Chosen NR = $g(h(NR_j))$

21   **End**

22   **End**

23   **End**

27   **If** NR's energy when connected with GW < 0,

28   **Then** remove from the routing path

28   **End**

29   Update the neighbor energy information

30   **Termination Criteria**

31              $P(l(z, X) \leq r) = 0.$

32   **End**

33 Return $g(h(NR_j))$

---

congested and packets start dropping and get lost. But with a rational nanonetwork employed with learning elements, congested routes can be identified and avoided by observing the aforementioned QoI attributes. It can respond to undesired scenarios proactively, by routing the data through a different path consisting of R_4, and R_8, as shown in Figure 5.3(b).

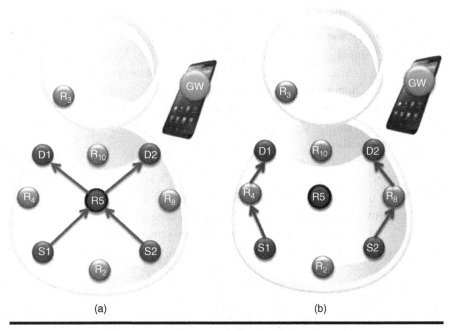

Figure 5.3 (a) **Typical routing and** (b) **rational routing in the IoNT.**

5.4.2 Reasoning

In the RDDA algorithm, we also employ the reasoning element for more rational nanonetworks. A modified version of the analytic hierarchy process (AHP) [31] is considered in order to implement this element of cognition in the IoNT. The reason we choose AHP for reasoning is that it supports multiple-criteria decision making while deciding on which path to deliver. For example, if we have imbalance in selecting the next hop for data delivery in energy-constrained nanonetwork, the set of NRs which provides the lowest energy consumption while satisfying the fairness attribute will be chosen even though it might degrade other metrics such as the network delay or cost. This means fairness and energy are prioritized over cost and delay in this situation. If two alternative paths can guarantee the same in terms of fairness and energy, then the next attribute to compare will be delayed, followed by the cost. The assumed AHP algorithm provides a method for pairwise comparison of each of the aforementioned attributes and helps to choose the node that can provide the best network performance in the long run. Example 2 elaborates more on this.

Example 2. Assume a three-level hierarchy in the AHP: objective, attribute, and alternatives as shown in Figure 5.4. A fundamental scale for pairwise comparisons is then used to set priorities for the nanonetwork attributes at the CNRs. Given the very limited energy constraint in IoNT, we would assign the highest priority to lifetime (or energy), followed by fairness and then delay. We tabulate

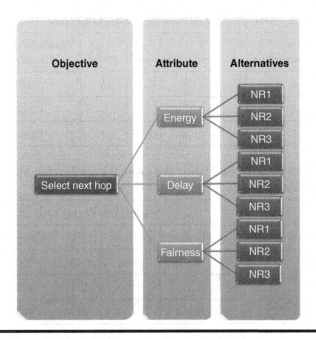

Figure 5.4 The hierarchical structure used in AHP algorithm.

the relative priorities of these attributes using pairwise comparisons and generate Table 5.2. From Table 5.2 we derive Table 5.3. Then, we apply the following steps:

1. Represent the values of Table 5.3 in an n × n matrix A, where n is the count of considered QoI attributes.
2. Get the Eigen vector of A,
3. Separate the absolute value of the Eigen vector,
4. Decide the best alternative according to most appropriate (i.e., highest) value in the Eigen vector as depicted in Table 5.4.

Accordingly, the modified AHP steps in prioritizing the nanonetwork QoI attributes is described in Algorithm 5.2.

Table 5.2 Pairwise Comparison of the Nanonetwork QoI Attributes

Lifetime	4	Fairness	1
Lifetime	6	Delay	1
Fairness	3	Delay	1

Table 5.3 AHP for QoI Attributes versus Objective

Objective - Best Attribute	Lifetime	Fairness	Delay	Relative Priorities of the Attributes
Lifetime	1	4	6	0.691
Fairness	1/4	1	3	0.2176
Delay	1/6	1/3	1	0.0914

Table 5.4 AHP Evaluating the Overall Priorities for All Possible NRs

Best Candidate for Next Hop NRx	Priority With Respect To			
	Lifetime	Fairness	Delay	Objective
NR_1	0.252	0.015	0.101	0.375
NR_2	0.2	0.018	0.11	0.329
NR_3	0.164	0.019	0.116	0.296

Algorithm 5.2 Modified AHP Algorithm

1   **Function AHP (P)**
2   **Input**
3   P: End-user defined priorities for the desired QoI attributes.
4   **Output**
5   NR_x: Next-hop $NR_x \in \{NR1 ..... NR_n\}$ with best P
6   **Begin**
7   **Initialize:** priority matrix for traffic type; Success=0;
8   **While** $P(l(z,X) \le r) > 0$
9   AHP_analysis(Next-hop NRs vs. attributes)
10   Next hop $NR = NR_x$
11   Transmit data to next-hop NR
12   **If** (next hop == GW)
13   *Success=1;*
14   **Else**
15   *Choose next-hop NR*
16   go to step 8
17   **End**
18   **If** *(Success==0)*
19   GW sends the request again.
20   **End**
21 **End**

5.5 Performance Evaluation

In this section, we evaluate the performance of the proposed RDDA against the two routing categories of the nanorouting approaches in the literature, namely the SPA and NNA algorithms. Based on the aforementioned system models, we summarize these two baselines' as follows.

5.5.1 Shortest Path Algorithm (SPA)

As we mentioned before, SPA is one of the most dominant categories in nanorouting. In this study, it has been used under two conditions: (1) polling condition, when the GW makes a request, and (2) pushing condition, when a data packet is to be sent from a nanosensor to the GW. In polling, the GW chooses randomly an NR and adds its index to the forwarded packet. Then from that NR, the packet is transmitted to the next hop NR based on the shortest path to the destination. On the other hand, the aim of the pushing is to transmit data packets to the nearest NR that is on the shortest path to the GW. This routing strategy is briefed in Algorithm 5.3.

Algorithm 5.3 SPA Approach

---

1   **Function SPA**
2   **Input**
3   An NR
4   **Output**
5   NR chosen by SPA to deliver data towards GW
6   **Begin**
7   **For** each NR, Hop count = 0;
8   Identify an NR to start the route.
9   List all neighboring NRs from NR.
10   **If** there is at least one NR_j connected with this NR
11   **For** each source NR_j
12              $f(NR_j) = $ Distance to GW
13   **Select** $g(NR_j) = min(f(NR_j))$
14   **Else**
15   There is no NR.
16   **End**
17 **End**
18 **Return** $g(NR_j)$

5.5.2 Nearest Neighbor Algorithm (NNA)

NNA is the second category for nanorouting and it is considered as an improved version of the SPA approach. NNA can be briefed in Algorithm 5.4, where it calculates the shortest path based on the four-direction strategy as discussed in Section 5.2.

In the two aforementioned baselines, SPA represents a straightforward approach in cutting down unnecessary energy consumption while choosing the shortest path. And thus, it has been chosen as one of the competitive alternatives in the area. On the other hand, the NNA algorithm follows the shortest path based on the available common communication channels between the heterogeneous nodes. It chooses the nearest neighboring NR, which has a non-diagonal connection with the nanosensor. Unlike the SPA approach, this one takes into consideration the heterogeneity of the IoNT and utilizes more capable nodes rather than just the nanosensor nodes. Consequently, NNA increases the hop count a bit so that it can save more energy, and hence, prolong the network life. The neighboring NRs' energy level and distance from GW are compared in RDDA. It chooses NR, which meets the requirements for transmission. In addition, it provides more lifetime to network. Thus, we compare our proposed RDDA with both NNA and SPA in this research. A detailed description of our experimental setup is given in the following section.

Algorithm 5.4   NNA Approach

---

1   **Function NNA**

2   **Input**

3   An NR

4   **Output**

5   NR chosen by NNA to deliver data towards GW

6   **Begin**

7   **For** each NR, Hop count = 0;

8   Identify an NR to start the route.

9   List all neighboring NRs from NR.

10   **If** there is at least one NR_j connected with this NR

11   **For** each NR_j

12      $f(NR_j)$ = Distance to GW

13   **Select** $g(NR_j) = max(f(NR_j))$

14   **Else**

15   There is no source NR.

16   **End**

17   **End**

18   **Return** $g(NR_j)$

5.5.3 Experimental Setup

In this section, a detailed description of the performed experiment for validation and verification purposes has been introduced. In addition to computer-based simulations, real sensing/relaying devices have been used in a test-bed for practical verifications. The test-bed is designed for the proposed IoNT architecture, and it assures the coherence functionality of the system and the integration of different IoNT components in terms of their reliability and lifetime. Actual deployment in the laboratory has been performed in order to evaluate the performance of the aforementioned routing algorithms. The use of TI CC2530 programmable motes [32] has enabled a fine-tuned programming, with a more sophisticated base for carrying out tasks that require high-end processors and devices. We used the JDK interface that is provided by Sun for programming the CC2530 nodes. These nodes have the ability to associate with multiple resources and offer relaying capabilities. Meanwhile, the nanosensor device is composed of the following basic components: a sensing unit, a central processing unit (CPU), a communication subsystem, and an optional storage unit. The processing unit executes the system software in charge of coordinating, sensing, and routing tasks and is interfaced with a storage unit. A communication subsystem interfaces the device to the network and is composed of a transceiver unit and communication software. The latter includes a communication protocol stack and proposed system software/algorithms. Finally, the whole system is powered by a power unit that is supported by an energy scavenging feature. In order to limit our search space, we assume a virtual grid, where SNs are placed on the grid vertices. We assume up to 1500 total SNs communicate with one GW via 36 NRs.

We used NS3 as simulation tool for this purpose as well. The simulation is processed in three platforms, which are Windows, Linux, and OSX, for validation purposes. We executed our simulation 100 times for each experiment and plotted the average results. More details about our simulation are summarized in Table 5.5.

Table 5.5   Simulation Parameters and Values Based on [26]

Parameter	Value
Target area	100nm × 100nm
Number of nodes	SNs: 1500, NRs: 36, GW: 1
Communication Range	SN: *130nm*, NR: *200nm*, GW: *400nm*
Initial Energy	SN: *31000J*, NR: *110000J*, GW: *Unlimited*
Energy Consumption	SN and NR (Receiving): *31.2 uJ/bit* SN and NR (Transmitting): *53.8 uJ/bit*

5.5.4 *Threats to Validity*

In this subsection, we elaborate more on practical aspects related to IoNT para-digms. Since this type of paradigm is still under investigation, there are several threats to be addressed in any proposed solution. First of all, the recommended/ used system models can be inaccurate and lead to significant failures in practice. Accordingly, a test-bed has been designed and utilized in this study. This test-bed has been examined under operational conditions, such as dense deployment areas and multi-hop short-range communication links with interference effects. Moreover, the targeted performance metrics have been selected carefully to match with the IoNT requirements. For instance, energy is one of the critical aspects in any nanonetwork, and thus, energy savings have been checked and analyzed in comparison to typical approaches in the literature via both simulations and experimental work. It is worth pointing out also that the selected performance metrics can be inadequate unless such an experimental work and test-bed have been performed and developed in order to observe the accurate system behavior in practice. The test-bed runs a client application that remotely connects to the BS through the IoNT infrastructure. This interface gives the freedom to issue queries to the network, retrieve data from the storage, and assign new metrics via remote access. And due to complexity of the targeted IoNT system, the suggested components have been designed in a modular fashion for easy replacement and development.

5.5.5 *Performance Metrics and Parameters*

In this research, we assess our proposed RDDA in terms of seven main metrics:

- *Path length (hops):* the total number of hops while packet is carried from source SN to sink (GW).
- *Network lifetime (rounds):* the number of transmission rounds until the closest NRs to GW is dead.
- *Remaining energy (joule):* the energy level at the end of network lifetime in the NRs and SNs.
- *Repair time:* the required time by an NR to repair a route/path failure that occurs when an intermediate node does not receive the data packet after a few attempts.
- *Fairness:* the ratio (percentage) of NRs which have been fairly treated while selecting the next hop. This metric shows the adeptness of distributing the load across the nanonetwork.
- *Latency:* the delay time from event occurrence until a desired outcome is achieved, such as storing a temperature value on the local GW memory after being retrieved from the corresponding data source.

Also, we assess our proposed RDDA in terms of two main parameters:

- ***GW request time (seconds):*** represents the average time interval (amount) between two consecutive requests sent from the same GW.
- ***Request time (seconds):*** the time that the request is made periodically.
- ***Transmission rounds:*** the average number of transmission rounds between nodes during network lifetime.

5.5.6 Simulation Results

According to our previous discussion, we assume, as seen from Figure 5.5, that we have 36 NRs and 1 GW in an area of 1000nm × 1000nm. Each RN is linked to a set of adjacent RNs, which are connected in all directions. In Figure 5.5, we can see small and big circles denoting the NRs' and GW's range, respectively, where the GW can scope any NR which exists in its circle. Moreover, the connections

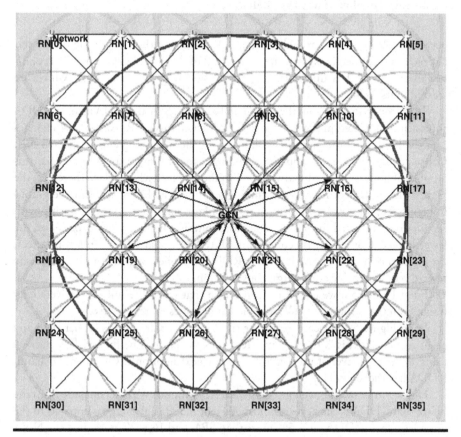

Figure 5.5   Assumed nanonetwork topology.

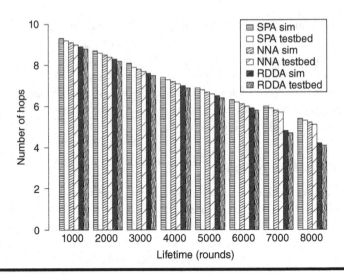

Figure 5.6 Number of hops vs. lifetime.

between GW and NRs are bidirectional. Accordingly, the GW in Figure 5.5 has bidirectional connections with the closest NR14, NR15, NR20 and NR21; additionally, the GW has unidirectional connection with the other NRs which are located in its vicinity.

The comparison between the two main baselines, SPA and NNA against RDDA approach, is depicted in Figure 5.6. Paths which connect NRs to the GW changes over the lifespan of the nanonetwork based on the remaining energy in each NR. Consequently, the average hop count changes, and it's important to note that this number of hops is proportionally related to the average packet delay. Therefore, the higher the number of hop count, the higher the delay. Inferring from the graphs in Figure 5.6, we observe that the number of hops has an inverse relationship with the lifetime measured in rounds. The more rounds a nanonetwork experience, the fewer hops it should use. By comparing SPA, NNA, and RDDA in all instances, the number of hops is always less when RDDA is applied, which is a desired feature toward reducing delay and energy consumption.

From Figure 5.6 we can also deduce also that when lifetime is relatively long (more than 6000 rounds), RDDA approach becomes more effective.

From Figure 5.7, we can clearly see that the latency for SPA is higher than any other approach. This is because it determines the path to be followed by packets using the energy level of the nearest NRs (or next hop) only. On the other hand, RDDA has the least delay.

Figure 5.8 shows the average energy levels per round. We can conclude from the figure that RDDA is the best and most efficient in terms of energy compared to the other baselines. RDDA saves 5% more than SPA, and 16.85% more than the NNA algorithm in terms of energy.

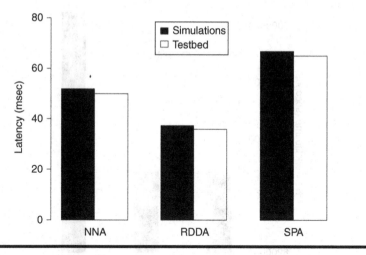

Figure 5.7 Comparison of latency.

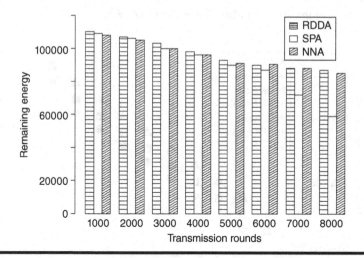

Figure 5.8 Comparison of average energy level at NRs vs. transmission rounds.

Figure 5.9 shows the lifetime of the network for the three baseline algorithms. The X-axis shows the different types of the network algorithms, while the Y-axis shows the lifetime in seconds. From the graph, we observe that RDDA has extended lifetime, while NNA and SPA achieve almost the same lifetime span. RDDA exceeds the other two algorithms by 562100 seconds.

Figure 5.10 shows how the lifetime of the system is affected by the requests made by the GW. The X-axis shows the request time in seconds, while the Y-axis shows the network lifetime in seconds. From the figure, we can see that the lifetime of the system increases with the increase of request time. Obviously, the lifetime

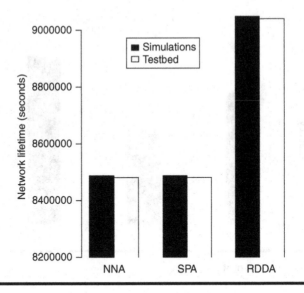

Figure 5.9 Comparison of network lifetime (seconds).

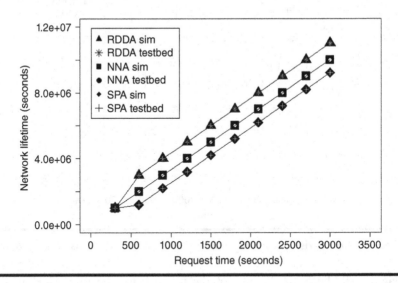

Figure 5.10 Comparison of network lifetime according to GW's request time.

of both SPA and NNA are less than the achieved lifetime by RDDA. Hence, we conclude that the increase in request time of the GW can increase the lifetime of the nanonetwork.

Figure 5.11 depicts the comparison of a one-hop energy level from the GW. The X-axis shows the GW nearby NRs, while the Y-axis shows the energy level of a given NRs. NR[14], NR[15], NR[20], and NR[21] are chosen specifically because

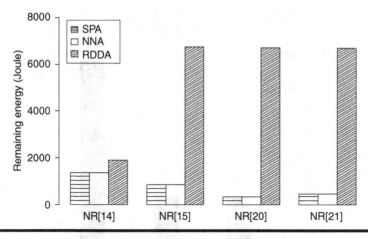

Figure 5.11   Comparison of one-hop NRs' energy level.

they have bidirectional connection with the GW and will be the most stressed nodes in the nanonetwork as they are the closest to the GW and must be used in any communication with it. Comparing the energy levels under the three different routing approaches, we notice that although the energy levels for NNA and SPA are the same in NR[14], NR[15], NR[20], and NR[21], RDDA is different and outperforms all of them. And thus, RDDA increases the network lifetime, and it is much better in terms of energy saving.

Figure 5.12 depicts the comparison of the three approaches in terms of the transmission rounds' count. From this figure, we observe that RDDA outperforms both NNA and SPA which confirms its efficiency in terms of prolonging the nanonetwork lifetime. In Figure 5.13 we examine the fairness level in each of the applied routing approaches. We define fairness as the ability of the system to echo the energy exhaustion rate at each nanosensor by redistributing the load over all the available NRs. That is because the IoNT paradigm is supposed to handle multiple resources under different restrictions and energy constraints. In this figure, we observe that the fairness of RDDA is more than that of the other two alternatives. The main reason for these results is that RDDA uses the aforementioned learning and reasoning elements in deciding the next hop, and hence, the processed requests are evenly distributed among the available NRs.

Meanwhile, route repair occurs when an NR tries to repair a route/path failure that occurs when an intermediate node does not receive the data packet after a few attempts. NR tries to repair the route by examining its routing table for alternate data forwarding points. Figure 5.14 shows the average route repair time which has been experienced by all of the three routing approaches. We observe that SPA has the highest route repair time, followed by NNA. And on the other hand, RDDA has the least route repair time.

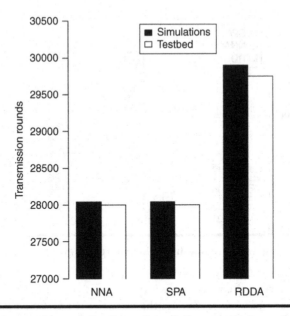

Figure 5.12 Comparison of the three data delivery techniques based on total number of transmissions.

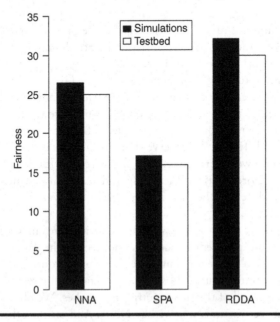

Figure 5.13 Comparison of the routing approach fairness.

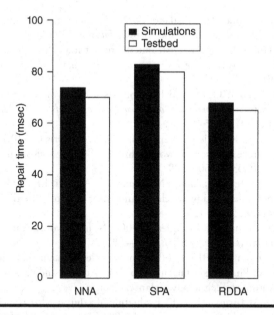

Figure 5.14   Comparison of the ability to repair per routing approach.

5.6 Conclusions

In this research, we examined two different categories of routing methods in the IoNT paradigm, namely the SPA and NNA, in terms of energy consumption, delay, and fairness. We proposed the RDDA approach, which is unique in prolonging the nanonetwork lifetime without violating other QoI attributes in IoNT. We concluded that RDDA can save a significant amount of energy. Additionally, this approach reduces the number of hop counts by roughly 23%. This is a significant achievement, especially when we learn that nanonetwork lifetime has an inverse relationship with the hop count. Furthermore, we explained how the hop count can be used to show on-spot and average delays at NRs. Moreover, we demonstrated how the RDDA approach provided the longest network lifetime. Even though both SPA and NNA demonstrate that they are efficient in terms of transmission rounds and energy consumption, the general results show that RDDA outperforms both of them.

References

1. NSTC/NNI. July 10, 2012. Nano.gov [online]. Available: http://www.nano.gov/node/847.
2. Chang Y., MIT. June 19, 2008. MIT tech review [online]. Available: www.technologyreview.com/s/410293/smarter-faster-nano-sensor/.

3. Al-Turjman F., Imran M., Vasilakos A. 2017. Value-based caching in information-centric wireless body area networks. *Sensors Journal*, 17(1), 1–19.

4. Ali S., Madani S. July 2011. Distributed efficient multi hop clustering protocol for mobile sensor networks. *The International Arab Journal of Information Technology*, 8(3), 302–309.

5. Al-Turjman F. 2017. Optimized hexagon-based deployment for large-scale ubiquitous sensor networks. *Springer's Journal of Network and Systems Management*, DOI: 10.1007/s10922-017-9415-2.

6. Nikolidakis S.A., Kandris D., Vergados D.D., Douligeris C. 2013. Energy efficient routing in wireless sensor networks through balanced clustering. *Algorithms*, 6, 29–42. DOI: 10.3390/a6010029.

7. Pierobon M., Jornet J.M., Akkari N., Almasri S., Akyildiz I.F. 2014. A routing framework for energy harvesting wireless nanosensor networks in the Terahertz Band. *Wireless Networks*, 20(5), 1169–1183.

8. Akyildiz I.F., Brunetti F., Blázquez C. 2008. Nanonetworks: A new communication paradigm. *Comput. Netw.*, 52(12), 2260–2279.

9. Jornet J.M. December 2013. Fundamentals of electromagnetic nanonetworks in the terahertz band. Ph.D. dissertation, School of Electrical and Computer Engineering, Georgia Institute of Technology, Atlanta, Georgia.

10. Hasan M.Z., Al-Turjman F. 2017. Evaluation of a duty-cycled asynchronous X-MAC protocol for vehicular sensor networks. *EURASIP Journal on Wireless Communications and Networking*, DOI: 10.1186/s13638-017-0882-7.

11. Hasan M.Z., Al-Rizzo H., Al-Turjman F. 2017. A survey on multipath routing protocols for QoS assurances in real-time multimedia wireless sensor networks. *IEEE Communications Surveys and Tutorials*, DOI: 10.1109/COMST.2017.2661201.

12. Hasan M.Z., Al-Turjman F., Al-Rizzo H. 2017. Optimized multi-constrained quality-of-service multipath routing approach for multimedia sensor networks. *IEEE Sensors Journal*, 17(7), 2298–2309.

13. Samet H. February 2008. K-nearest neighbor finding using MaxNearestDist. *IEEE Transactions on Pattern Analysis and Machine Intelligence*. 30(2), 243–252.

14. Al-Turjman F. 2016. Cognition in information-centric sensor networks for IoT applications: An overview. *Springer Annals of Telecommunications Journal*, 1–16. DOI: 10.1007/s12243-016-0533-8.

15. Al-Turjman F. 2017. Cognitive-node architecture and a deployment strategy for the future sensor networks. *Springer Mobile Networks and Applications*, DOI: 10.1007/s11036-017-0891-0.

16. Liaskos C., and Tsioliaridou A., Ioannidis S., Kantartzis N., Pitsillides A. 2016. A deployable routing system for nanonetworks. *IEEE Int. Conf. on Comm. (ICC)*, 1–6.

17. Tsioliaridou A., Liaskos C., Ioannidis S., Pitsillides A. 2015. Corona: A coordinate and routing system for nanonetworks. *Proc. of the Second Annual Int. Conf. on Nanoscale Computing and Comm.*, ser. NANOCOM '15, New York, NY, ACM, 18:1–18:6.

18. Yu H., Ng B., Seah W.K.G. 2015. Forwarding schemes for EM-based wireless nanosensor networks in the terahertz band. *Proc. of the Second Annual Int. Conf. on Nanoscale Computing and Comm.*, 17:1–17:6.

19. Zhou R., Li Z., Wu C., Williamson C. 2012. Buddy routing: A routing paradigm for nanonets based on physical layer network coding. *2012 21st Int. Conf. on Computer Comm. and Networks (ICCCN)*, 1–7.

20. Tsioliaridou A., Liaskos C., Ioannidis S., Pitsillides A. 2015. CORONA: A coordinate and routing system for nanonetworks. Proceedings of the Second Annual International Conference on Nanoscale Computing and Communication, ACM. Boston, MA.
21. Singh G., Al-Turjman F. 2016. A data delivery framework for cognitive information-centric sensor networks in smart outdoor monitoring. *Elsevier Computer Communications Journal*, 74(1). 38–51.
22. Liaskos C., Tsioliaridou A. 2015. A promise of realizable, ultra-scalable communications at nano-scale: A multi-modal nano-machine architecture. *IEEE Transactions on Computers*, 64.5, 1282–1295.
23. Al-Turjman F. 2017. Cognitive caching for the future fog networking. *Elsevier Pervasive and Mobile Computing*, DOI. 10.1016/j.pmcj.2017.06.004.
24. Agoulmine N., Kim K., Kim S., Rim T., Lee J.-S., Meyyappan M. 2012. Enabling communication and cooperation in bio-nanosensor networks: Toward innovative healthcare solutions. *IEEE Wireless Communications*, 19(5), 42–51.
25. Al-Turjman F., Hassanein H., Ibnkahla M., January 2013. Quantifying connectivity in wireless sensor networks with grid-based deployments. *Elsevier Journal of Network & Computer Applications*, 36(1), 368–377.
26. Akyildiz I.F., Jornet, J.M. December 2010. The internet of nano-things. *IEEE Wireless Commun.*, 17(6), 58–63.
27. Al-Turjman F. 2018. *Wireless sensor networks: Deployment strategies for outdoor monitoring.* New York: Taylor and Francis, CRC. ISBN 9780815375814.
28. Jornet J.M., Akyildiz I.F. October 2011. Channel modeling and capacity analysis for electromagnetic wireless nanonetworks in the terahertz band. *IEEE Trans. Wireless Commun.*, 10(10), 3211–3221.
29. Hang Y., Ng B., Seah W. 2015. Forwarding schemes for EM-based wireless nanosensor networks in the terahertz band. Proceedings of the 2nd Annual International Conference on Nanoscale Computing and Communication. ACM.
30. Al-Turjman F. 2017. Information-centric sensor networks for cognitive IoT: an overview. *Annals of Telecommunications*, 72(1), 3–18.
31. Singh G., Al-Turjman F. 2016. Learning data delivery paths in QoI-aware information-centric sensor networks, *IEEE Internet of Things Journal*, 3(4), 572–580.
32. Texas Instruments CC2530 [online]: http://www.ti.com/product/CC2530.

Chapter 6

A Value-Based Caching Approach for WBAN

Fadi Al-Turjman

Department of Computer Engineering, Antalya Bilim University, Antalya, Turkey

Contents

6.1 Introduction 88
6.2 Related Work 90
    6.2.1 Location-Based Caching...91
    6.2.2 Content-Based Caching...92
    6.2.3 Node Functionality-Based Caching 93
6.3 System Models 94
    6.3.1 ICN-Based WBAN Model 94
    6.3.2 Delay Model 95
    6.3.3 Age Model 96
    6.3.4 Popularity of On-Demand Requests 97
    6.3.5 Channel Communication Model 97
6.4 VoI Cache Replacement 98
    6.4.1 Theoretical Delay Analysis 99
6.5 Performance Evaluation 100
    6.5.1 Performance Metrics..101
    6.5.2 Simulation Parameters 101
    6.5.3 Simulation and Results 102
6.6 Conclusions..109
References 109

6.1 Introduction

Great interest is currently invested in wireless body area networks (WBANs) used for remotely monitoring and reporting users' health vitals [1]. Composed of various sensors, both wearable and implanted, a WBAN relies on wireless connectivity to transfer the collected data to an Internet-based data management service [2, 3]. Applications span healthcare, sports, and industry, and have received attention from various commercial and civilian sectors. A healthcare example is that of remotely monitoring a patient's condition, allowing patients to maintain lifestyles and comfort while offsetting hospital management costs. Such a setup would facilitate giving early warnings to patients with heart or neurological problems, and providing a heads up for nearby caregivers. Similar systems can be applied for airplane pilots and long-route drivers to alert them to their awareness level. WBANs can also be fitted to workers in harsh environments, such as mines, to avert serious consequences from physical or chemical strain. A typical system comprises the following components: (1) a fitted WBAN; (2) a gateway between the WBAN and the Internet; (3) a service for data transfer over the Internet; and (4) a service for data processing, providing interface and alerts to the caregivers. For a successful and reliable WBAN operation, a harmonious nearby set of data resources needs to be simultaneously engaged in all components. The human body is a challenging terrain for wireless communication, especially given the WBAN power constraints (due to limitations on the amount of energy allowed to penetrate the human body). Accommodating various levels of harshness in the surroundings in terms of temperature, dust, humidity, etc., dictates a resilience requirement to an expanded set of failure possibilities that includes partial or complete failure of the WBAN nodes and reduced levels of activity or accuracy as batteries deplete, which form serious threats for losing critical in-network data before it is utilized. Equally, a WBAN needs to maintain high integrity for the data collected. This requirement stresses both the integrity of in-network storage elements and the computations of the collected data at the different components. Moreover, a fitted WBAN needs to sustain different levels of mobility. Since the WBAN elements rely on each other to gather and process data, mobility may be temporarily or permanently detrimental to the network operation by breaking some functional communication links that affect the in-network data retrieval.

Hence, nodes and links are prone to several risks leading to high probabilities of failures, and several nodes in the WBAN may become disconnected. We are characterizing such circumstances by the probability of node failure (PNF) and probability of link failure (PLF). At the same time, network elements rely on instantaneous data updates in order to maintain healthy user conditions. Delivery and accessibility of in-network data can, hence, be extremely challenging, especially when dealing with highly variable data such as neurological signals, which necessitates timely recording. Thus, delay/latency in accessing data forms a critical design factor in these networks. Meanwhile, the priority of different collected data depends on the nature of the monitored phenomena and the criticality of the user's condition. Therefore, a prioritization that can be either preset or reactive to the user's condition at both the WBAN level and during transfer to the caregivers is needed. Consequently, for a successful and reliable operation and monitoring of the WBAN system, the information-centric network (ICN) paradigm shall be applied in such systems for immediate in-network data access.

ICN is the next-generation model for the Internet that can cope with the user's requests/inquiries regardless of their data-hosts' locations and/or nature [4]. The current Internet model is suffering from the exchange of huge amounts of data while still relying on the very basic network resources and IP-based protocols. Meanwhile, ICNs promise to overcome major communication issues related to the massive amounts of distributed data on the Internet. ICNs adopt a content-centric architecture which focuses more on the networked data itself rather than the meta-data. Luckily, the architecture of these ICNs closely match with the emerging communication trend that aims at exchanging big data over tiny and energy-limited WBANs in order to realize numerous attractive projects such as the smart-planet and the Internet of Things [5–7].

In order to enable WBAN to support this trend in communication and function as a reliable platform, we proposed the cognitive framework in our previous work [8]. In [8], an information-centric scheme is proposed for typical wireless sensor networks (WSNs) using cognitive (intelligent) in-network devices that make dynamic routing decisions based on specific knowledge and reasoning observations in the WSNs. Knowledge representation using the *<attribute, value>* pair, and reasoning using analytic hierarchy process (AHP) techniques are employed by the cognitive device in order to decide on the best data route. AHP can be applied on quality of information (QoI) attributes in next-generation WBANs such as reliability, delay, and network throughput observed over the communication links/paths [9,10]. The QoI is defined in this work as the level of satisfaction experienced/perceived by the end-user about the received information from the WBAN. Attributes such as reliability, latency, and throughput are used to evaluate the QoI. It is worth pointing out here that QoI is not referring to the quality of service (QoS) term used in typical WSNs [3]. In fact, QoS takes care of the operational aspects of the network, while QoI is associated with the characteristics of the sensory information made available to the end user/sink node.

Cognitive WBAN is able to significantly outperform the non-cognitive WBAN paradigms [11]. However, this cognitive WBAN framework necessitates a reliable in-network caching feature. In this chapter, we propose the use of a value of sensed information (VoI) cache replacement strategy. It identifies the most suitable data to be replaced in order to maintain prolonged data availability periods. Where VoI is defined as the value of information that is going to be exchanged in the WBAN and it consists of the information age, popularity, delay, and energy depletion cost.

Unluckily, traditional cache replacement strategies in the literature [12,13] have been designed mainly for IP-based computer networks and data-centers, which have distinct characteristics in locating data from the envisioned next-generation networks such as the lightweight WBANs. In fact, choosing the most appropriate caching strategy can have significant implications on the overall network performance in terms of the data publisher's load, hit ratio, and time-to-hit metrics. Several works in the literature have addressed each of these metrics separately. However, since a single WBAN can serve numerous kinds of applications/users with varying design requirements, we believe in the necessity of a generic dynamic utility function that can consider all the aforementioned metrics while setting different weights for each depending on the WBAN application.

To this end, we provide a novel utility function that sets a value to each cached data item in a cognitive WBAN framework. We provide a cache replacement strategy that depends on the VoI in choosing the most appropriate data to be replaced in the cache. We compare our VoI-based strategy against three other significant approaches in the literature, which we call least recently used (LRU), first in first out (FIFO), and least value first (LVF), while considering varying parameters in ICN-WBAN, including the cache size and requests' counts. Moreover, we testify our caching approach under flat and hierarchical storage structures. In the flat structure, we assume only one level of caching. However, in the hierarchical one, we assume two levels of caching.

The rest of the sections in this chapter are organized as follows: A literature review on caching approaches in ICN-based WBANs is provided in Section 6.2. In Section 6.3, we introduce our WBAN-specific system model, based on which we build our proposed caching strategy. Section 6.4 explains the proposed VoI approach and utility function. Section 6.5 presents extensive simulation results of the VoI in comparison to other caching strategies. And finally, we conclude our work in Section 6.6.

6.2 Related Work

WBANs can be viewed as a special type of content-oriented wireless sensor network (WSN), and, as such, they inherit some of a WSNs' general characteristics. The uniqueness of WBANs as a paradigm, however, mainly stems from dealing with the terrain of the human body. The human body, as a communication terrain, poses unusual challenges for both in-body and in-air communication. Many factors, including body mobility, changes in posture, size, weight, or water content, in addition to coexistence of different technologies in the same band, affect the throughput of the WBAN. For life-threatening cases, for instance, the resulting data packet loss/miss is simply unaffordable. Moreover, essential considerations for safety and comfort prohibit the use of capable transmission, sensing, and processing techniques [1, 2]. A substantial component, as well, of the WBAN operation involves processing some measured information at the network level, i.e., before delivering it to the gateway. Leading motivators for in-network caching include energy savings, reduced delay, measurement aggregation (summary, average, etc.), measurement verification, etc. [14]. In general, given the nature of the WBAN application, there is a high demand for operational reliability in terms of data availability and accessibility. While such considerations do not dismiss the plethora of work with WSNs in the literature, certain challenges that are unique to WBANs call for novel solutions. Within the context of WBANs, there is an apparent void in addressing the challenges in data caching, which has always been treated as a key feature in ICNs. In this chapter, we propose the use of ICN-based solutions in addressing this void in WBANs in order to augment and provide more in-network data hits in a timely manner.

At the core of ICNs, data have to "live" near their requesters. This is the task of caching schemes. Thus, efficient caching mandates two properties: (1) ensuring an updated copy of the requested data resides at entities close to the region of interest (in terms of reachability latency); and (2) that copy remains "live" for as long as interest in it exists. Caching is coupled with routing and naming architectures. For example, in data-oriented network architecture (DONA), the coupling of naming tuples enables in-network caching for any entity in the network that could hold a valid copy. Architectures differ in deciding which entity is allowed a copy of the data and the basis upon which each entity would retain the data. A recent effort in age-based caching argues for a twofold metric for caching an NDO replica. If it resides at a network edge, or has higher popularity, it will remain cached for a longer duration. Entities which hold replicas will collaborate in "tuning" the age counter to manifest such factors. A core disadvantage at many caching protocols is the inherent need for bookkeeping. The resulting message-exchange overhead cannot scale to the Internet and remain efficient. We are bound to analyze caching schemes under the following conditions: (i) communication overhead per NDO; (ii) storage requirements; (iii) NDOs with different priorities; and (iv) granularity in assessing request frequency, types, locations, etc. We review the different caching techniques in ICNs and identify the techniques that are best suited for the caching decisions in WBANs. The in-network caching in ICN-based WBANs can be categorized into the following categories: (A) location-based caching; (B) content-based caching; and (C) functionality-based caching.

6.2.1 Location-Based Caching

Chai et al. in [15] have argued against caching the data everywhere in ICNs and recommended caching less in order to achieve better network performance. Their caching policy claims that data shall be only cached at the nodes having the highest probability of getting a cache hit on the data delivery path. Eum et al. [16] have proposed a content-oriented network (CON) architecture, called cache aware target identification (CATT). This architecture assumes a topology-aware caching policy, where a node on a downloading path is selected for caching as long as it has the highest connectivity degree based on the geographical location of the node.

However, this kind of node can form a geographical bottleneck in the network. Meanwhile, authors in [17] have investigated the performance of topology-based replica placement on Internet router-level topology and found that the router-level fan-out placement is almost as good as that of the greedy placement of the replica. Moreover, they found that a fan-out based replica placement method needs to be carefully designed to be efficient in content-oriented architectures.

Works in [15–17] are based on the node degree or node fan-out for replica placement, but these methods cannot be universal because node degree-based solutions cannot be good solutions if most of the nodes have a similar, relatively low degree or fan-out. Bhattacharjee et al. [18] have considered the use of various self-organizing or active cache management strategies in which nodes make globally consistent decisions about caching and revealed that in many cases, these self-organizing caching schemes yield better average delays than traditional approaches (cache at transit nodes), using much smaller per-node caches. A selective neighbor caching approach which selects an appropriate subset of neighboring proxies that minimize the mobility costs in terms of expected average delay and caching costs has been proposed in [19]. This approach is based on proactively caching data requests and the corresponding metadata to a subset of proxies that are one hop away from the proxy. This kind of caching usually follows a FIFO technique in replacing the cached data based on its in-network locality.

6.2.2 Content-Based Caching

The main objective of the work in [20] is to minimize the Internet service provider (ISP) traffic and accessed in-network devices by caching frequently requested data at ISP-specific routers. The main problem addressed here is to provide effective caching strategies for these routers to coordinate their data replacement based on their content. Guided by optimal replica placement, the authors have presented two popularity-based caching algorithms. However, this work may not be practical as authors have assumed only one gateway in an ISP network.

Cho et al. [21] have proposed a content caching approach in which the cache size is adjusted based on data popularity. In this approach, an upstream node recommends the number of chunks to be cached at its downstream node, which increases exponentially as the data requests increase in order to reduce communication and cache management overhead. It distributes content chunks towards the network edge (from where the data requests come) considering the content popularity and distance relation. However, the different sizes of data chunks have not been considered in this reference. An age-based distributed cache approach aiming at reducing the data publisher load and in-network delay for ICNs has been proposed in [22]. This approach provides a lightweight cooperative mechanism to control where data contents' ages are dynamically updated implicitly. It spreads popular contents toward the edge of the ICN and, meanwhile, eliminates the unnecessary replicas at the intermediate ICN nodes. Yet this approach suffers from maintaining highly dynamic contents, and thus, nodes which are far away from the server may experience long time periods to refresh their contents. This caching category can be represented by the famous term "least recently used" (LRU) as it replaces the outdated data based on the content itself rather than its locality.

6.2.3 Node Functionality-Based Caching

To unleash the full potential of ICNs, the role (function) of the in-network caching node shall be taken into consideration and will consider which content to cache at the management/control level rather than guessing it at the data level. Authors in [23] remarked on the side effects of delegating the caching decision to the data level and proposed a specific approach to handle data caching at the control level. The proposed approach can be tuned to create a balance between the benefits and costs overhead. However, it is only applicable at a small scale, and it may not accommodate the massive amounts of data contents in the Internet. In [24], authors investigated the trade-off between caching data contents in a distributed IP-based network and the new emerging ICN architectures such as the content-centric network (CCN). They applied their study on the real traffic mix resulting from several functional resources such as the Web, file sharing, and multimedia streaming. It has been demonstrated that caching videos in routers offers more cache hits. Nevertheless, the other types of content would likely be more efficiently handled in very large-capacity storage devices in the core of the network. Thus, this kind of caching is not efficient in ICNs, and it follows the FIFO techniques in caching.

In this chapter, although we need to use a content-centric approach, the traditional cache replacement approaches cannot be applied to an ICN-based WBAN. This is because of the unique resource constraints of the sensor network, the uncertainty of the wireless medium, and the need to be aware of application-specific requirements in ICN architectures. The resource limitations of the sensor nodes include limited power supply, storage space, and heterogeneity in terms of the sensors used and the node functions. In addition, the same content (sensed data) cannot be replicated into multiple caches without associating them with location, because the sensed information may be different in different parts of the network, and it may change over time too, which is unlike the case of ICNs. This makes cache replacement trickier in information-centric sensor networks. In addition, the replacement policy should take into account the type of user/application requests coming to the network, the sensor node availability at different locations (as nodes eventually die out), and also the sensing duration for different sensors on board the sensor nodes.

Consequently, the Internet has progressed towards more information-centric paradigms, where the focus is on delivering named blocks of data to users at the network edge rather than establishing end-to-end connections to the web server. So the design of the cache replacement policy in ICNs must be a dynamic one, based on the user's request trends and the application on hand. In [25], we proposed the concept of dynamic caching via assigning dynamic values for the exchanged in-network contents. Then we chose the least-valued ones to be replaced. Accordingly, this technique is called the LVF approach. However, in LVF there was no cognition in making the caching decisions, and it assumed the traditional flat storage structure (i.e., one level of caching). In this chapter, and unlike other related work [26, 27], dynamic caching decisions are made based on specific knowledge and reasoning observations in the network. Based on these cognitive elements/observations, we prioritize between the cached contents in a hierarchical storage system (i.e., in a multilevel caching). Accordingly, we provide a novel utility function that sets a value to each data item based on application-specific metrics, such as required quality of communication channel, delay, and data age. This makes our proposed VoI approach able to cope with the next-generation sensor networks trend in communication.

6.3 System Models

In this part of our chapter, we elaborate on our ICN-based WBAN model in addition to its corresponding data popularity, age, and delay models.

6.3.1 ICN-Based WBAN Model

The main components in our ICN network model consist of sensor nodes (SNs) to sense the environment and capture physical changes, and relay nodes (RNs) or local cognitive nodes (LCNs) to which the SNs report these changes. They interact with the RNs and LCNs as shown in Figure 6.1. LCNs have elements of cognition, e.g. knowledge, reasoning, and learning, which help in interpreting common requests and queries' responses. They interact with SNs, RNs, and the sink. RNs forward the data received from SNs to the sink or any neighboring local cognitive nodes in response to received requests from the user. The sink is where all the collected data is delivered. The sink node is also enhanced with cognitive elements to be more intelligent in managing the network performance based on data traffic type, and it is called a global cognitive node (GCN).

The type of data traffic that an ICN-based WBAN paradigm can handle is categorized into one of the following: (1) Type I: On-Demand; (2) Type II: Periodic; and (3) Type III: Emergency. Each of these types is associated with a different QoI value on the cached data, based on the WBAN application. We select the network reliability (R), latency (L), energy (E), and throughput (T) as the four main attributes, whose combined value decides the QoI of the cached data. We are not using the absolute values for these attributes. Instead, we associate priorities with each of these attributes for every request type and make these priorities decide the importance of the absolute value of the attributes, as shown in Table 6.1. For instance, number 1 indicates the highest priority and number 3 indicates the lowest. The 'x'

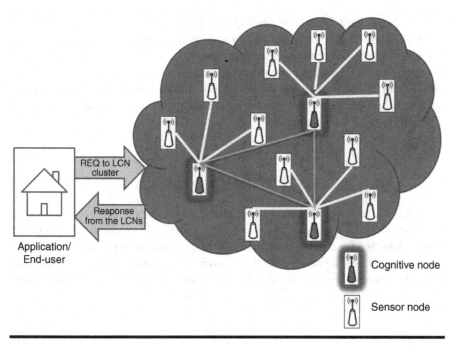

Figure 6.1   Network model with sensor nodes and cognitive nodes.

Table 6.1   Energy and Quality of Information (QoI) Attribute Priority for Different Data Traffic Types

Quality of Information (QoI) — Attributes				
Request Type	*Latency (L)*	*Energy (E)*	*Reliability (R)*	*Throughput (T)*
Type I: On-Demand	x	3	1	2
Type II: Periodic	1	2	4	3
Type III: Emergency	1	1	x	2

in Table 6.1 indicates a "do not care" condition. This means that there are no strict requirements on the value of the QoI marked with an 'x,' and its value does not affect the caching decision.

6.3.2 Delay Model

Different sensors have different durations for which they need to be exposed to the environment so that they can capture the sensed readings accurately. This affects the duration of the on-time of the sensor node, which in turn affects the lifetime of the sensor node [28]. In order to prolong the lifetime of the sensor node, it is useful to store the sensed data for longer when the delay involved in acquiring the reading is longer. This is called the sensing delay. In addition, if data has to be propagated from sensor nodes to LCNs every time data is requested, it would add to the propagation delay of the data, especially if the sensor nodes are located far away from the sink. Thus, the delay components we consider are the sensing delay δ and the propagation delay τ. Accordingly,

$$\tau \propto n, \tag{6.1}$$

$$\delta \propto \max(d_1, d_2, d_3, ...d_k) \tag{6.2}$$

where k is the total number of sensors available on board of the sensor node, and d_i represents the fixed sensing delay value of the sensor type i (see Table 6.1). Thus, the sensing delay is a function of the maximum delay from among the sensor types that have been activated to provide fresh data. Putting these two delays together, the total delay (Δ) involved in delivering freshly sensed data to the sink is a combination of the sensing and propagation delay, given by Eq. (6.3).

$$\Delta = \tau + \delta \tag{6.3}$$

6.3.3 Age Model

Our age model makes use of the following two conditions to decide what content should be dropped from the cache. The first is based on the periodicity of the periodic request (Type 1 traffic), and the second is based on when the node's cache is full. We make use of the periodicity of the periodic request, because freshly sensed data has to be provided at the start of each periodic request cycle. Thus, when the cache is full at the end of one periodic request cycle, old data can be discarded from the cache. The age of a sensed attribute-value pair is represented by its time-to-live (TTL) which is based on the periodicity of the request of each application type. This value is provided to the LCN by the GCN/sink. Since we are not considering the use of historic data, our model implies that cached contents may be refreshed after every periodic time interval, as long as the data is being transmitted to the sink at the end of each cycle.

$$TTL_{Si} \propto T_{periodic} \tag{6.4}$$

Eq. (6.4) represents that the TTL of the sensed information (Si), represented as an attribute-value pair, is directly dependent on the periodicity of a request in Type 1 traffic flow. In case the application requires that the periodic data be stored for a prolonged duration of time, for example 24 hours, before making a single transmission to the sink, the cache retention period becomes a function of the transmission cycle's periodicity.

6.3.4 Popularity of On-Demand Requests

Traffic flow generated in response to on-demand requests have been classified as Type 2 traffic. More users may be interested in a particular type of sensed data, or a specific sensed data may be requested more times by one or more users. Such sensor data is said to be popular, and can be retained for longer in the LCN's cache. Thus, the popularity of the sensed attribute-value pair is given by Eq. (6.5).

$$Popularity_{Si} \propto \mathrm{Re}\, q_{Si} \,/\, \mathrm{Re}\, q_{total} \tag{6.5}$$

where $\mathrm{Re}\, q_{Si}$ is the total count of requests for an attribute-value pair received at an LCN, and $\mathrm{Re}\, q_{total}$ is the total count of requests received by that LCN within a particular round of an ICN-based WBAN operation. In addition, when sensor nodes start to die out in the network, LCNs should store the data for longer to maintain their availability. When the primary LCN storing such data itself starts to die out, storing the data in neighboring LCNs provides extra storage guarantees and ensures availability of data in the network for longer. This storage requirement based on non-availability of alive sensor nodes is managed by the planning algorithm for data delivery based on the traffic flow in the network and remaining energy at LCNs.

6.3.5 Channel Communication Model

Here we elaborate on the assumed channel model in our wireless communications [29]. The transmission power utilized by the ICN-based WBAN nodes is represented as T_{po} and the transmission range between the *BS* and an *SN* is represented as T_r. The expression of the channel model can be provided as

$$C_M = A\rho T_{P_0} T_r^{-\alpha} \tag{6.6}$$

where C_M is the transmission power of the *BS*, A is the constant gain factor for power provided by the antenna, and amplifier gain, ρ, is the small-scale constant for fading factor, and α is the path loss exponent. The transmission range between the *WBAN* nodes of i and j is denoted as $R_{ij}\left(i, j = 1, 2, 3, \ldots\ldots, N\right)$. The range for *WBAN* nodes (i and j) and the *BS* is denoted as r_i and the power transmission for node i is defined as p_i. The link interference is expressed as

$$I_{BS} = \frac{A\rho T_{P_0} T_r^{-\alpha}}{A\rho T_{P_i} r_i^{-\alpha}} \tag{6.7}$$

6.4 VoI Cache Replacement

For cache replacement in ICNs, we need to ensure that we choose data appropriately for storage based on the following criteria: Firstly, data that takes longer to sense should be stored for longer to conserve the sensor node's energy. Secondly, data storage must be a function of the periodicity of the requests based on the traffic type. This will help to store data until fresher data is available and to service requests for different traffic types in a timely manner. Lastly, the value of information based on its age. Hence, the freshness of data is also an important criterion when servicing requests for data on demand. Since these criteria are known and fixed, the cache replacement plan can be programmed into the LCN.

We propose a VoI-based cache replacement strategy for the LCNs in an ICN. Our cache replacement approach adopts the aforementioned system models to achieve an efficient cache management strategy that can handle the following three types of content:

1. Delay-based content: The delay sensitivity of the cached data content is a measure specified by the requesting user to indicate how long the consumer is willing to wait for it. Examples of delay-sensitive data can be found in applications serving areas of emergencies (e.g. disaster or health emergency).
2. Age-based content: Some contents are more sensitive to aging. For instance, if a user requests information about the traffic updates for the coming 30 minutes, then any related content that does not cover this time interval is useless.
3. Demand-based content: This is a measure of the data popularity which is specified by the frequency of requesting specific data.

Accordingly, the VoI-cache management approach employs three parameters to set a *value VoI$_{Si}$* per sensor node S_i reading. This value is dependent on the history of the data content within each operational round. At the beginning of each round and based on the aforementioned models, every content resets its *VoI$_{Si}$* value according to the following function:

$$VoI_{Si} = \alpha \times \Delta + \beta \times TTL_{Si} + \gamma \times Popularity_{Si} + \lambda \times 1 / I_{BS} \qquad (6.8)$$

where α, β, γ, and λ are the tuning parameters that are specified based on the traffic type and the user requests. The strength of the VoI is mainly in its ability to prioritize based on the targeted WBAN application. Given the importance of the data popularity factor, especially in disaster cases, we would like to remark that it is recommended not to separate this factor totally from the proposed VoI concept. However, it is complementary with other important factors such as the age, delay, and cost in terms of the consumed energy and caused interference within the WBAN. Thus, to improve the basic priority caching method, the weights of each of the parameters can be adjusted to find the most efficient approach. The delay

Algorithm 6.1   Drop Least *Vol*_{si}

1	**Function Vol** (*content*)
2	**Input**
3	*content: A content item within the ICN.*
4	**Begin**
5	**for** each node, **do**
6	**for** each duty round, **do**
7	Set *value* of each *Vol*_{Si} in the cache based on Equation (8)
8	**if** *cache_full*
9	Check history of the data requests
10	Drop the data content of the least *Vol*_{Si}
11	**End if**
12	**End for**
13	**End for**
14	**End**

sensitivity parameter serves in assuring the least delay. The popularity parameter is important as it takes into consideration the most frequently demanded data packets. The packet age value is also vital since it considers packets in the cache that have not been used for a long time and replaces them with more relevant data. In the following, Algorithm 6.1 provides the steps to be executed by each node if its cache is full to drop data with the least Vol_{Si}.

6.4.1 Theoretical Delay Analysis

One of the key objectives in VoI is to minimize the worst delay experienced between any sensor-node pair. Thus, we adopt the delay model introduced in [28]. Most importantly, we assume its discretized delay metric, which can be tuned to achieve any desired accuracy. Due to dense network topologies formulated in WBAN, a relatively long multi-hop path can easily exist between the source node and the corresponding destination. Consequently, the delay components we are considering in this case are the transmission/processing delay ψ represented by the number of hops multiplied by ψ, and the propagation delay based on the speed of signal and the Euclidian distance between the two ends (source and destination). The latter delay is extremely dependent on the speed of the link (or signal speed) and the Euclidian distance between the source and destination. It varies based on the utilized technology and its corresponding standards and the transmission medium.

Accordingly, the experienced delay in WBAN can be described as follows: We define a delay step ω which is the distance a wireless signal would travel in one time unit. Assuming E_{ij} is the Euclidian distance between a source node *i* and a destination node j, then the discrete propagation delay over a single-hop link (i, j) would be $\left[\frac{E_{ij}}{\omega}\right]$. Hence, the discrete delay over a multi-hop path is the sum of the discrete delays of single-hop links that constitute that path. Note that ω (and the time unit) can be made small enough to meet any desired accuracy. Thus, the upper-bound delay for a single hop (D_{single}) and multi-hop (D_{total}), can be respectively defined as follows:

$$D_{single} = \left[\frac{E_{ij}}{\omega} + \psi\right] \tag{6.9}$$

and

$$D_{total} = \sum_{total\ hops} \left[\frac{E_{ij}}{\omega} + \psi\right] \tag{6.10}$$

6.5 Performance Evaluation

In this section, we provide initial performance evaluation results for the VoI-based cache replacement technique, which we have compared with FIFO, LRU, and LVF techniques using NS3 (https://www.nsnam.org/overview/media-kit/), a discrete event simulator. We adopt the aforementioned delay, popularity, and data-aging models proposed in Section 6.3, which are described in Eqs. (6.3), (6.5), and (6.7). In our simulation, r is set to 142 mm, P is set to 3 dBm, and data rate is set to 25 kbps. The packet size is 512 bits. Every WBAN node has an initial energy of 50 J and generates 5 pkts/s. Our simulations involve networks with 40, 60, 80, 100, 120, or 140 nodes randomly deployed in a field of radius R that ranges from 400 mm to 1400 mm. Thus, mesh topologies are formulated and assumed in this simulation. For each network size, we test 20 instances and take the average. To generate a trajectory for the requested data, we use the directed diffusion (DD) approach. As for the arbitrary variables specified in our VoI caching function in Eq. (6.8), we set $\alpha = 0.2$, $\beta = 0.2$, $\gamma = 0.3$, and $\lambda = 0.3$.

We make use of the cache hit ratio to compare the performance of the different cache replacement strategies. Cache hit ratio is defined as the ratio of the number of times requested data was found in the cache divided by the total number of times data was requested from the cache. The storage cache is implemented as a single storage level in one case (L1 cache) and as a hierarchy of two storage levels in another case (L1 and L2 caches). Simulation results are compared for VoI, LVF, LRU, and FIFO replacement techniques. These simulations were run at cache sizes ranging from 10 to 100 Mbytes, and the simulations end after serving 1000 packet requests. There are 100 different requests from which the packet requests are randomly generated.

6.5.1 Performance Metrics

To compare the performance of the proposed VoI approach, we track ICN-specific metrics to achieve qualitative conclusions for the targeted in-network caching problem. We simulate the performance of an ICN-based WBAN with the detailed physical layer NS3 built-in parameters so that we achieve realistic simulation instances. The four considered performance metrics are as follows:

■ Cache hit ratio is simply the fraction of time a request arrives at a node to which that cache is attached but does not contain the requested data item. It is the average hitting ratio over all the in-network caches. We preferred to look at average time to hit data and hitting ratio more than publisher load, but we generally expect publisher load to improve as the other metrics improve as well.

■ Time-to-hit data (TTH) is found by simply logging all the total costs of the request and response paths incurred by every sensor node. Ideally, ICN-based WBAN is supposed to minimize the total average time-to-hit data per request.

■ In-network latency (delay) represents the end-to-end delay as described above. Note that we differentiate between latency to hit data and in-network latency since the two metrics may differ because of mobility or disruption conditions

■ Average request per publisher (ARP) is measured in number of data requests per hour (req/h) and it represents the average load per publisher in an ICN paradigm. We track publisher load by monitoring the total fraction of data requests that had to be satisfied by a data publisher.

6.5.2 Simulation Parameters

Many of the ICN paradigm parameters have to remain fixed while our simulation instances are generated. In particular, the parameters of our simulation are as follows:

■ Percentage of nodes with caches (PoC): This parameter is our primary method for controlling the extent of caching in our ICN. By varying this parameter, we can study the sensitivity of metrics like time-to-hit-data to the caching extent.

■ Connectivity level (degree): This represents how tightly connected the ICN-based WBAN is. We use the connectivity matrix, based on our described communication model in Section 6.3.

■ Data popularity: This indicates how frequently a specific data content is requested. This metric is measured in percentage with respect to other requested data contents. This parameter is represented by a single Poisson process parameter in order to give the content replacements per time unit.

■ PNF (%): This is the probability of a physical damage and/or a battery depletion for the deployed WBAN node due to harsh operational conditions. This parameter is chosen to reflect the impact in case of disaster scenarios or fragmented WBAN.

6.5.3 Simulation and Results

The following figures depict the achieved results. Our first objective is to confirm that increasing the extent of caching in ICNs, in terms of both size and number of levels, will reduce time-to-meet data for all cache policies.

According to Figure 6.2, we can deduce that the value of sensed Information is not efficient in a level one cache; however, LVF, FIFO, and LRU have better performance in this case. The performance gains do not increase much under the VoI approach when the cache size is increased beyond 30, due to the limited number of caching levels. Since the decision making is taken only at cognitive nodes (CNs), implementing the VoI-based cache replacement strategy necessitates more resources to achieve better cache hit ratios. This cannot be achieved by only implementing the first level of cache, where the proposed prioritization method will not be effective simply because there are no alternatives to prioritize between them.

In Figure 6.3, where we have two levels of caching (one on LCN and the other on RN), we find that the VoI surpasses the other two replacement techniques. The advantage offered here by the VoI-based cache replacement technique is that it can replace data based on the user requirements. However, in other cache replacement techniques such as LVF, FIFO, and LRU, authors are only worried about the match of the data request packet numbers, irrespective of the age of data, its popularity, or the delay involved in sensing and transmitting it to the sink. However, with the VoI-based technique, factors like age of data, popularity, and interference are also considered during the data replacement. Thus, older, easy-to-get data is replaced by fresher data that are much harder to get. Accordingly, we suggest the use of two

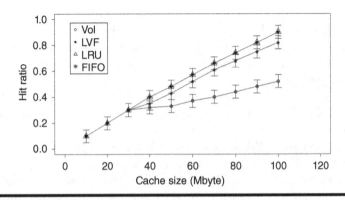

Figure 6.2 Cache size versus the hit ratio with one-level caching.

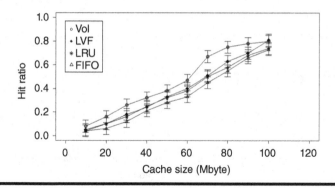

Figure 6.3   Cache size versus the hit ratio with two-level caching.

levels of caches, one at the CN and one at RN. Since we have only one type of the considered requests, there is a maximum performance gain when the cache size is increased beyond 60 Mbyte. This supports our claimed VoI approach against the heterogeneous types of requests in WBAN applications.

The figures for the next set of simulations (Figures 6.4 and 6.5) are set to analyze the performance of the cache replacement strategies as the number of requests that a given network needs to serve increases from 500 to 5000 req/h. The cache size is set to be 100 Mbyte and the number of request types are fixed at four.

As seen in the above figures, the advantage offered by the value of sensed information-based cache replacement technique is that it can replace data based on the user requirements and VoI of data. Other cache replacement techniques are only worried about the match of the data request packet numbers, irrespective of the age of data or its popularity or the delay involved in sensing and transmitting it to the sink. However, with the VoI-based technique, factors like age of data, popularity, and VoI are also considered during replacement. Older data is replaced by fresher data unlike other techniques that look only for a number match irrespective of its age.

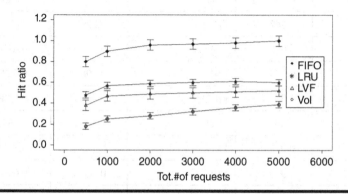

Figure 6.4   Total number of requests versus the hit ratio with one-level caching.

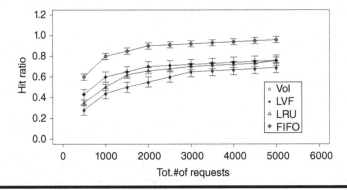

Figure 6.5   Total number of requests versus the hit ratio with two-level caching.

Based on this, we suggest the use of two levels of caches, one at the LCN and one at RN. At the LCN, we can use the VoI-based cache replacement strategy, and at RN, we can use either the FIFO or LRU to make the computation less complex. The size of first-level cache and second-level cache is set be 100 Mbyte and number of packet requests is set to be 10,000.

From Table 6.2, we can see that for the two levels of caches, the best possible combinations are: VoI-based replacement strategy at L1 cache and VoI- or LRU-based replacement strategy at L2 cache. Although FIFO-based techniques sometimes perform better in terms of hit ratio, we cannot say that they meet the user requirements in terms of the age of the data and delay requirements. Since the decision making is only at LCNs, implementing the VoI-based cache replacement

Table 6.2   Two-Level Caching Comparison

L1 Caching Policy	L1 Hit Ratio	L2 Caching Policy	L2 Hit Ratio	Tot. Hit Ratio
VoI	0.811542	LRU	0.81743	0.81542
VoI	0.547	FIFO	0.7792	0.6194
VoI	0.899	VoI	0.0099	0.81743
LRU	0.754	VoI	0.398	0.6837
LRU	0.9	FIFO	0	0.71818
LRU	0.802	LRU	0.49	0.75125
FIFO	0.9	LRU	0	0.71818
FIFO	0.9	VoI	0	0.61818
FIFO	0.9	FIFO	0	0.65125

strategy at LCNs can significantly help the network in saving more resources if a cache hit is found at the first level of cache.

Figures 6.6 and 6.7, below, represent the findings from the simulation experiment of the two-level cache. From both figures, the extent of cache availability increases proportionally. According to Figure 6.6, we observe that the overall time-to-meet data, which is our main performance metric, is reduced in all performance policies. However, the VoI policy performs best with higher proportions of nodes attached to caches. On the contrary, Figure 6.7 shows that there is an increase in the data hit for all the approaches, and, hence, we conclude that the VoI is better due to its ability to replace the most relevant data according to an ICN-specific set of attributes.

In Figures 6.8–6.10, connectivity level (degree) is the examined parameter. From Figure 6.8, we deduce that there is an increase in time-to-hit data as the

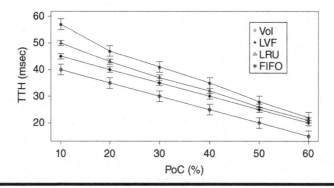

Figure 6.6 Time-to-hit ratio versus the percentage of nodes with caches in a randomly generated information-centric network (ICN)-based wireless body area network (WBAN).

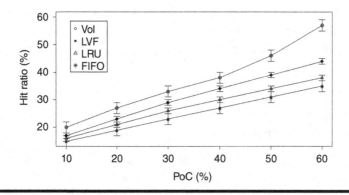

Figure 6.7 Hit-ratio versus the percentage of nodes with caches in a randomly generated ICN-based WBAN.

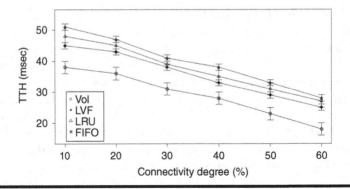

Figure 6.8 Time-to-hit ratio versus the connectivity degree percentage.

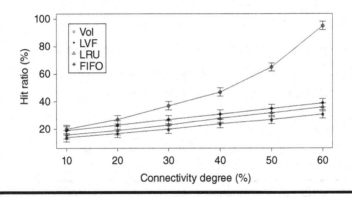

Figure 6.9 Hit ratio versus the connectivity degree percentage.

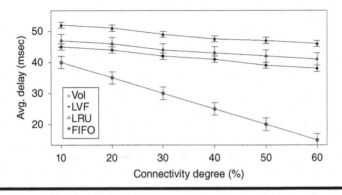

Figure 6.10 Average in-network delay versus the connectivity degree percentage.

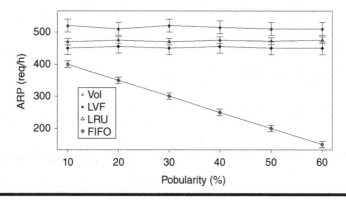

Figure 6.11 Publisher load versus the data popularity.

ICN-based WBAN connectivity increases in all the approaches. However, we notice that VoI is less dependent on the network and, hence, better than the other two approaches. VoI is more dependent on the data type of highly desired property in the ICN-based network. Figure 6.9 shows the data hit performance against a varying network connectivity degree, and while applying the VoI scheme, we notice that the data hit increases exponentially while the network connectivity increases. Nevertheless, the data hit of the other two approaches increases linearly. Moreover, Figure 6.10 shows that VoI is the best in terms of delay. This can be attributed to the application of the delay factor while deciding what data to replace. Figure 6.11 shows the effect of data popularity in terms of publisher load. The VoI tops LRU and FIFO as the popularity metric increases. This is a very desirable property in ICNs.

The VoI-based technique would be suitable for the ICN-based approach, as we propose to use a named data association for the sensed data, such as attribute-value pairs, and the cache size can be decided based on the different types of user requests that the network is expecting to serve. We can expect that the users are more satisfied with the response received from the LCN because it implements the caching policy based on both, data popularity and energy involved in doing so while the network scales up to larger sizes.

Furthermore, we examined the four caching approaches, LVF, LRU, FIFO and VoI, in terms of the average publisher load (Figure 6.12) and average delay (Figure 6.13) impacts while considering disaster scenarios and/or fragmented WBAN, where failure of a critical node partitions the network into disjointed segments [30]. Based on Figure 6.12, we notice a sever effect on the average request per publisher (ARP) while the PNF is increasing. That is because all approaches are experiencing an exponential increase in the publisher load as the network becomes disconnected. However, using the proposed VoI approach, the increment is going linear, which can be a very desirable feature in WBANs,

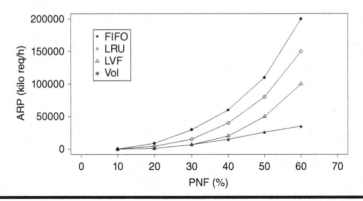

Figure 6.12 Publisher load versus the probability of node failure in the network.

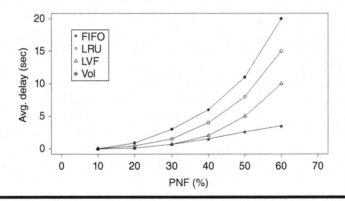

Figure 6.13 Average delay versus the probability of node failure in the network.

while experiencing harsh operational conditions and sever mobility effects. Moreover, we examined the VoI approach against the other three baselines while varying the PNF values from 10% to 60% to check the effect on the average data delivery delay in Figure 6.13. Again, we can see the exponential increase in delay while applying the other three approaches against the VoI-based approach. This can be returned to the special care that has been taken while caching/removing data by assigning the highest tuning parameter value to the popularity factor. This can help a lot in disaster and emergency scenarios. It is worth mentioning, however, that through the displayed results, the LVF approach always comes as a second best to VoI, with a considerable gap between the two. We explain this as a result of LVF being the closest caching approach to VoI in terms of treating data/information as measurable entities that can be compared and replaced accordingly.

6.6 Conclusions

The VoI-based technique is suitable for use in ICNs which use named data association for the sensed data and are expected to support node mobility in the future. We showed how the users are more satisfied with the response received from the proposed LCN components. These new components can retain information in their cache based on data popularity and various other parameters that can affect the process of data gathering and energy involved in doing so. Further, the VoI cache replacement strategy will help in graceful degradation of the network, as cached data can be provided from LCNs even after sensor node deaths. Moreover, extensive simulations to evaluate the impact of a disjointed (fragmented) network on the average delay and data publisher load have been studied to show the effectiveness of the VoI cache replacement strategy.

References

1. Fortino G., Galzarano S., Gravina R., Li W. 2015. A framework for collaborative computing and multi-sensor data fusion in body sensor networks. *Inf. Fusion*, 22, 50–70.
2. Fortino G., Giannantonio R., Gravina R., Kuryloski P., Jafari R. 2013. Enabling effective programming and flexible management of efficient body sensor network applications. *IEEE Trans. Hum. Mach. Syst.* 43, 115–133.
3. Al-Turjman F. October 13–14 2016. Impact of user's habits on smartphones' sensors: An overview. In *Proceedings of the HONET-ICT International IEEE Symposium*, Kyrenia, Cyprus, 70–74.
4. Singh G., Al-Turjman F. 2016. Learning data delivery paths in QoI-aware information-centric sensor networks. *IEEE Internet Things J.*, 3, 572–580.
5. SmartSantander. Future Internet Research and Experimentation. Available online: http://www.smartsantander.eu/ (accessed on 13 January 2017).
6. IBM|A Smarter Planet|Smarter Cities. Available online: http://www.ibm.com/smarterplanet/us/en/smarter_cities (accessed on 13 Dec. 2016).
7. Al-Turjman F. 2016. Cognition in information-centric sensor networks for IoT applications: An overview. *Ann. Telecommun.*, 1–16, doi:10.1007/s12243-016-0533-8.
8. Singh G.T., Al-Turjman F.M. 2016. A data delivery framework for cognitive information-centric sensor networks in smart outdoor monitoring. *Elsevier Comput. Commun.*, 74, 38–51.
9. Al-Turjman F., Alfagih A., Alsalih W., Hassanein H. 2013. A delay-tolerant framework for integrated RSNs in IoT. *Elsevier Comput. Commun.*, 36, 998–1010.
10. Al-Turjman F., Hassanein H., Ibnkahla M. 2013. Efficient deployment of wireless sensor networks targeting environment monitoring applications. *Elsevier Comput. Commun.*, 36, 135–148.
11. Gravina R., Alinia, P., Ghasemzadeh H., Fortino G. 2017. Multi-sensor fusion in body sensor networks: State-of-the-art and research challenges. *Inf. Fusion*, 35, 68–80.

12. Chankhunthod A., Danzig P., Neerdaels C., Schwartz M., Worrell K. January 1996. A hierarchical internet object cache. In *Proceedings of the 1996 Annual Conference on USENIX Annual Technical Conference*, San Diego, CA, 22–26.

13. Al-Turjman F. 2017. Information-centric sensor networks for cognitive IoT: an overview. *Annals of Telecommunications*, 72(1), 3–18.

14. Al-Turjman F., Hassanein H. October 21–24 2013. Enhanced data delivery framework for dynamic Information-Centric Networks (ICNs). In *Proceedings of the IEEE Local Computer Networks (LCN)*, Sydney, Australia, 831–838.

15. Chai W.K., He D., Psaras I., Pavlou G. 2013. Cache "less for more" in information-centric networks (extended version). *Elsevier Comput. Commun.* 36, 758–770.

16. Eum S., Nakauchi K., Shoji Y., Nishinaga N., Murata M. 2012. CATT: Cache aware target identification for ICN. *IEEE Commun. Mag.* 50, 60–67.

17. Radoslavov P., Govindan R., Estrin, D. June 20–22 2001. Topology-informed internet replica placement. In *Proceedings of WCW'01: Web Caching and Content Distribution Workshop*, Boston, MA.

18. Bhattacharjee S., Calvert K.L., Zegura E.W. March 29–April 2 1998. Self-organizing wide-area network caches. In *Proceedings of the Seventeenth Annual Joint Conference of the IEEE Computer and Communications Societies*, San Francisco, CA, 752–757.

19. Vasilakos X., Siris V., Polyzos G., Pomonis M. August 17, 2012. Proactive selective neighbor caching for enhancing mobility support in information-centric networks. In *Proceedings of the ICN Workshop on Information-Centric Networking*, Helsinki, Finland. ACM: New York, NY, 61–66.

20. Li J., Wu H., Liu B., Wang X., Zhang Y., Dong L. October 29–30 2012. Popularity-driven coordinated caching in named data networking. In *Proceedings of the Eighth ACM/IEEE Symposium on Architectures for Networking and Communications Systems*, Austin, TX, 200–211.

21. Cho K., Lee M., Park K., Kwon T.T., Choi Y., Pack S. March 25–30, 2012. WAVE: Popularity-based and collaborative in-network caching for content-oriented networks. In *Proceedings of the 2012 IEEE Conference on Computer Communications Workshops (INFOCOM WKSHPS)*, Orlando, FL, 316–321.

22. Ming Z., Xu M., Wang D. August 4–7 2012. Age-based cooperative caching in information-centric networks. In *Proceedings of the 2012 IEEE Conference on Computer Communications Workshops (INFOCOM WKSHPS)*, Orlando, FL. 268–273.

23. Wang Y., Lee K., Venkataraman B., Shamanna R.L., Rhee, I., Yang S. March 25–30, 2012. Advertising cached contents in the control plane: Necessity and feasibility. In *Proceedings of the 2012 IEEE Conference on Computer Communications Workshops (INFOCOM WKSHPS)*, Orlando, FL.

24. Fricker C., Robert P., Roberts J., Sbihi N. March 25–30 2012. Impact of traffic mix on caching performance in a Content-Centric Network. In *Proceedings of the 2012 IEEE Conference on Computer Communications Workshops (INFOCOM WKSHPS)*, Orlando, pp. 310–315.

25. Al-Turjman F. 2018. Fog-based caching in software-defined information-centric networks. *Elsevier Computers & Electrical Engineering Journal*, 69(1), 54–67.

26. Vasilakos A.V., Li Z., Simon G., You W. 2015. Information centric network: Research challenges and opportunities. *J. Netw. Comput. Appl.* 52, 1–10.

27. Amadeo M., Campolo C., Quevedo J., Corujo D., Molinaro A., Iera A., Aguiar R.L., Vasilakos A.V. 2016. Information-centric networking for the Internet of Things: Challenges and opportunities. *IEEE Netw.* 30, 92–100.

28. Al-Turjman F., Hassanein H., Ibnkahla M. 2015. Towards prolonged lifetime for deployed WSNs in outdoor environment monitoring. *Elsevier Ad Hoc Netw. J.*, 24, 172–185.

29. Al-Turjman F., Hassanein H., Ibnkahla M. 2013. Quantifying connectivity in wireless sensor networks with grid-based deployments. *Elsevier J. Netw. Comput. Appl.*, 36, 368–377.

30. Alnuem M.A., Zafar N.A., Imran M., Fayed M., Ullah S. 2014. Formal specification and validation of a localized algorithm for segregation of critical/non-critical nodes in MAHSNs. *Int. J. Distrib. Sens. Netw. 2014*, 140973.

Chapter 7

Adaptive WBAN in the IoNT

Fadi Al-Turjman[1] and Hamed Osouli Tabrizi[2]

[1]Department of Computer Engineering, Antalya Bilim University, Antalya, Turkey
[2]Sustainable Environment and Energy Systems, Middle East Technical University Northern Cyprus Campus, Kalkanli, Güzelyurt, Turkey

Contents

7.1 Introduction 113
7.2 System Model 115
7.3 Packet-Size Optimization for a Batteryless WBASN 119
7.4 Numerical Results 120
    7.4.1 Packet Error Rate with Respect to Payload 120
    7.4.2 Energy Efficiency versus Payload Simulations 122
    7.4.3 Energy Efficiency with Respect to Distance 124
    7.4.4 Results for a Batteryless Network 124
7.5 Conclusion 126
References 126

7.1 Introduction

The wireless body area sensor network (WBASN) has recently gained attraction for its twenty-four-hour capability in chronic disease monitoring as well as fatal event prognostics such as heart attacks. WBASN is a heterogeneous network that can utilize sensor nodes which are attached or in close proximity to the body, on-body, or implanted inside the human body. Most sensor nodes proposed in the literature rely on batteries as their power supply, which limits their lifetime to a maximum of a few years and introduces battery replacing issues, especially for implanted sensors [1–5]. Recent trends in prolonging the lifetime of such a network employ energy harvesting methods to make the sensor nodes self-powered while utilizing body heat, body movement, RF or solar energy [6–10]. Body heat is proven to be a suitable candidate for energy harvesting towards eliminating the battery usage [11]. However, relying on the stringent energy budget of the energy harvesters imposes many challenges to the performance of the network. Energy efficiency is the most important concern in implementation of WBASN based on battery, while performance is the main objective to be optimized with the very limited amount of energy for infinite time.

Packet-size optimization is investigated as an aspect of energy efficiency for battery-powered WBASNs to improve battery lifetime [12–15]. Nevertheless, packet-size optimization for a fully batteryless WBASN has not been considered. In a batteryless network, dynamic packet-size optimization before each cycle of data transmission is crucial to prevent energy depletion of sensor nodes and therefore loss of vital data. Since the ambient energy level can vary rapidly, packet-size optimization should be done for every cycle of data transmission. In [12], an analytical method for packet-size optimization for a single-hop direct battery-powered WBASN is proposed for considering some error-control methods. The method is extended in [13], which includes a cooperative network. In [14], packet optimization for single-hop, cooperative, and two-hop WBSNs is studied. In [15], performance analysis of a battery-based network, which considers as well a small thermoelectric energy harvester has been investigated. However, the self-powered WBASN, which relies only on the harvested energy, is not considered. In such networks, packet size should be dynamically optimized regarding the amount of the power available at each node to avoid any transmission failure due to depletion of energy. In this chapter, a new method for dynamic packet-size optimization with respect to the amount of energy available at each node for each packet transmission is proposed. In addition, a scenario in which different sensor types can have different packet sizes is considered to take advantage of the network heterogeneity. Where the optimal packet size should be determined based on the available harvested energy between two consecutive cycles of a data transmission session. This in turn depends on the capability of the energy harvester. Even though energy efficiency is a crucial factor to extend battery life for WBASNs, it will not be the main criteria when considering fully batteryless networks, since energy harvesting provides infinity lifetime while on a very stringent energy budget. Instead, WBASNs should be optimized for the best performance considering the very limited but endless available harvested energy.

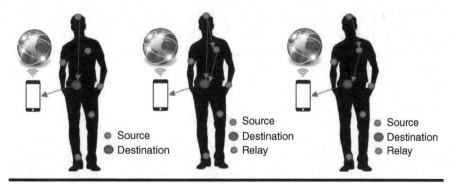

Figure 7.1 **(Left) Single-hop, (Middle) cooperative, and (Right) two-hop wireless body area sensor networks.**

The rest of the chapter is organized as follows. Section 7.2 discusses the system model. In Section 7.3, dynamic packet-size optimization problems constrained to the amount of harvested energy at each node are presented. Section 7.4 demonstrates the numerical results and discussion. Finally, the chapter ends with concluding remarks in Section 7.5.

7.2 System Model

In this chapter, we will consider three common network structures proposed in the literature for WBASN implementation. A single-hop implementation requires that sensor nodes would only communicate with the hub. In the cooperative network, the first data transmission is sent to a relay node which can be positioned between the hub and the source nodes. The relay node keeps the data for a specific time and if it overhears retransmission of the same data, it sends the data to the hub, increasing the probability of a successful packet transmission. In the two-hop structure, data is sent to the relay nodes and then to the hub. Figure 7.1 summarizes these three network structures.

In [12], an analytical model for energy efficiency of a single-hop direct WBASN is introduced based on the communication channel model, packet error rate, and energy consumption in each node. WBASN is a dynamic topology due to human body movement, and thus it varies the relative position of the sensor nodes. Therefore, in the path loss modeling, shading effect should be taken into consideration. Path loss is a parameter depending on the transmission channel as well as shadow effect, which can be described as

$$PL(d) = PL(d_0) + 10n \log_{10}\left(\frac{d}{d_0}\right) + S$$

where $PL(d_0)$ is the measured path loss at a reference distance d_0 from the transmitter, d is the distance in meter, n is the path loss exponent, and S is the loss due to shadow fading, which is normally distributed with a zero mean and standard deviation of δ_s. In addition, minimum energy requirements for a transmitter can be defined as follows:

$$P_{Rx} = P_{Tx} - PL$$

in which PL represents total power loss between a transmitter and a receiver and $P_{Rx}(min)$ represents the minimum power required by a receiver in order to receive a packet successfully. Minimum receiver power determines the signal to noise (SNR) at the receiver side. To have near-zero packet transmission error, the SNR ratio at the receiver should be addressed, which depends on the receiver circuit properties and noise level. The SNR ratio at the receiver can be defined as

$$\Psi(d) = P_{Rx} - Pn$$

where $\Psi(d)$ is the SNR ratio in dB of receiver and Pn is the noise power. Bit error rate (BER) increases when SNR increases. BER can be defined as following:

$$P_b^{OOK} = \frac{1}{2} erfc\left(\sqrt{\frac{1}{4}\frac{E_b}{N_0}}\right), \quad \frac{E_b}{N_0} = \Psi(d)\frac{B_N}{v_{TX}}$$

$$P_b^{BPSK} = \frac{1}{2} erfc\left(\sqrt{\frac{E_b}{N_0}}\right), \quad \frac{E_b}{N_0} = \Psi(d)\frac{B_N}{v_{TX}}$$

where P_b^{OOK} is the BER for on-off keying modulation scheme, P_b^{BPSK} is the BER for binary phase shift keying modulation scheme, B_N is the noise bandwidth and depends on the characteristics of the low pass filter on the receiver side, and v_{TX} is the data rate. Packet error rate is a function of bit error rate and the length of packet and can be defined as the following equation:

$$PER = 1 - (1 - BER)^l$$

where l is the packet length in bits. In order to keep the packet error rate close to zero, proper physical configuration and transmission conditions should be adjusted. Different bit error rate probabilities have been discussed in the literature with respect to modulation types and SNR ratios. Energy requirements for a single transmission include transmission and receiver circuit energy consumption in addition to the power transmitted by the antenna during the transmission period as follows:

$$E_{comm} = E_{TX} + E_{RX}$$

$$E_{TX} = E_{TX-elec} \cdot q + \frac{P_T}{v_{TX}} \cdot q$$

$$E_{RX} = E_{RX-elec} \cdot q$$

where E_{comm} is the total consumed energy per packet without considering packet retransmissions due to packet error rate, E_{TX} is the energy consumed by the transmitter node per packet, E_{RX} is the energy consumed by the receiver node per packet, $E_{TX-elec}$ is the energy per bit needed by transmitter electronics and signal processing, $E_{RX-elec}$ is the energy per bit needed by receiver electronics and signal processing, P_T represents the output transmit power which is the amount of energy spent in the RF amplifier, and q is the packet size which is equal to

$$q = s + h$$

where s is the payload and h is the packet overhead. Energy efficiency of a single-hop direct network can be defined as follows:

$$Eff = (1 - PER)\frac{x.(s+h)}{x.(s+h) + E_{RX,A} + E_{TX,A}}$$

$$x = E_{TX-elec} + \frac{P_T}{v_{TX}} + E_{RX-elec}$$

Energy efficiency for a cooperative network is given in [13].

For optimization, a cost function can be defined as follows without losing the generality of the equations.

$$CostF = -log(Eff)$$

CostF is the function that should be minimized with respect to s to get the optimum packet size. The optimization problem is shown as follows:

$$\hat{s} = arg \max_s Eff = arg \min_s CostF$$

In [13], similar method is applied for cooperative networks and the energy efficiency model is introduced as equation

$$Eff = (1 - PER_C)\frac{x.(s+h)}{E_b^c.(s+h) + E_{RX,A} + E_{TX,A}}$$

where PER_C is the packet error rate for the cooperative network and E_b^c is the energy consumption for a single-bit transmission. In a cooperative network, two independent scenarios lead to packet error: (i) unsuccessful transmission from source to destination and from source to relay node, or (ii) unsuccessful transmission from source to destination and successful transmission from source to relay while failure in transmission from relay to destination. Based on these scenarios, PER_C is given in equation

$$PER_C = PER_{SD} PER_{SR} + PER_{SD} (1 - PER_{SR}) PER_{RD}$$

Energy consumption for a single bit of data is also proposed considering a successful transmission from source to destination or successful transmission from source to relay and in the second time slot from relay to destination. Considering the energy consumption of each part E_1, E_2, and E_3, respectively, E_b^c can be written as equation

$$E_b^c = (1 - PER_{SD}) E_1 (s + h) + PER_{SD} PER_{SR} E_2 (s + h) + PER_{SD}$$
$$\cdot (1 - PER_{SR}) PER_{RD} E_3 (s + h)$$

$$E_1 = \left(E_{TX-elec} + \frac{P_T}{v_{TX}} + E_{RX-elec} \right)$$

$$E_2 = \left(E_{TX-elec} + \frac{P_T}{v_{TX}} + 2E_{RX-elec} \right)$$

$$E_3 = \left(2E_{TX-elec} + 3\frac{P_T}{v_{TX}} + 2E_{RX-elec} \right)$$

In [14], the energy efficiency model for the two-hop communication is introduced. Energy efficiency of two-hop network is given in equation

$$Eff = (1 - PER_{2-hop}) \frac{x.(s + h)}{E_b^{2-hop}.(s + h) + E_{RX,A} + E_{TX,A}}$$

Based on the two independent events that can happen in this network: (i) the unsuccessful transmission from source to relay and (ii) the successful transmission from source to relay with a failure in transmission from relay to destination, the packet error rate, PER_{2-hop}, and the energy consumption per each bit, E_b^{2-hop}, can be defined as depicted in the following equations:

$$PER_{2-hop} = PER_{SR} + (1 - PER_{SR}) PER_{RD}$$
$$E_b^{2-hop} = (1 - PER_{SR}) E_1 (s + h) + PER_{SR} \times 2E_1 (s + h)$$

7.3 Packet-Size Optimization for a Batteryless WBASN

Minimizing the *CostF* function with respect to *s* gives the optimum packet size. This optimization is done as part of the system design, since the battery provides a specific amount of energy to the system. However, in a batteryless solution, variable amount of energy will be harvested between every two packet transmissions, and therefore this optimization should be done for every packet transmission. The new constrained optimization can be defined as follows:

$$\hat{s} = arg \max_{s} \; Eff = arg \; \min_{s} \; CostF$$

$$Subject \; to \quad x.(s+h) + E_{RX,A} + E_{TX,A} \le E_{harvested}$$

This optimization problem can be solved using the result of a non-constrained optimization problem which has an analytical solution. Taking advantage of heterogeneity, it is assumed that each node can have a different packet size. The pseudocode for this optimization is stated as follows:

---

1 $\hat{s} = arg \max_{s} Eff = arg \min_{s} CostF$

2 $E_{harvested} = harvester\_power\_capacity \times \hat{s} \times Samplingrate \times bits \; in \; a \; sample$

3 $E_{Consumed} = (1+PER) \times \left(x.(s+h) + E_{RX,A} + E_{TX,A}\right)$

4 *if* $E_{harvested} - E_{Consumed} \ge 0$

   $\hat{s} = \hat{s}$

   else

   $\hat{s} = \hat{s} + 1$

   Go to 2

5 *End*

First, the non-constrained optimization method is solved and the result is used as the initial point to solve the constrained problem. Due to the nature of the body signals, their sampling rate is known. Next, harvestable energy within the time necessary for the packet to be measured can be calculated. Harvester power capacity reflects the capability of the energy harvester circuit and depends upon different energy harvesting and power-management circuits. In addition, it depends on the ambient condition and availability of the energy sources, for instance, temperature gradient for thermoelectric generators. In this chapter, power management circuit introduced in [15] is used as the energy harvesting unit. If the energy necessary to transmit the packet is more than the harvestable energy, then the node should wait longer so that enough energy can be harvested. This means the packet size will also increase due to the increase in time. Therefore, the algorithm increases the packet size by adding one more sample of data to it and does the calculations again till it reaches the point that harvested energy becomes enough for transmission of the packet.

7.4 Numerical Results

In this section, we provide numerical results and comparison of the energy efficiency criteria that is employed for battery-powered networks and performance as another figure of merit for batteryless networks. All simulation parameters are given in Table 7.1.

7.4.1 Packet Error Rate with Respect to Payload

Figure 7.2(a) shows the packet error rate with respect to payload for three different scenarios of one-hop communication. Since in-body communication can only be considered for distances less than 20 cm, packet error rate is considerably smaller. However, it is much higher for the no line of sight (NLOS) scenario and reaches maximum for a payload of 150 bits while it is around 300 bits for LOS. Figure 7.2(b) depicts the same axis for one-hop, cooperative, and two-hop

Table 7.1   Parameter Values

Parameter	Value
Overhead size	80 bits
ACK size	64 bits
Transmission power P_T	In-body: -10dBm
	On-body: -12dBm
Noise power P_n	-100 dBm
In-body channel model	d_0:5 cm, $PL(d_0)$: 49.81 dB, n: 4.22 δ_s: 6.81 dB, f: 402 to 405 MHz
On-body LOS channel model	d_0:10 cm, $PL(d_0)$: 35.2 dB, n: 3.11 δ_s: 6.1 dB, f: 2.45 GHz
On-body NLOS channel model	d_0:10 cm, $PL(d_0)$: 48.4 dB, n: 5.9 δ_s: 5.0 dB, f: 3.1 GHz
Data rate	In-body: 800 Kbps
	On-body: 2 Mbps
$E_{TX\text{-}elec}$	In-body: 18.75 nJ/bit
	On-body: 11.25 nJ/bit
$E_{TX\text{-}elec}$	In-body: 18.75 nJ/bit
	On-body: 11.25 nJ/bit

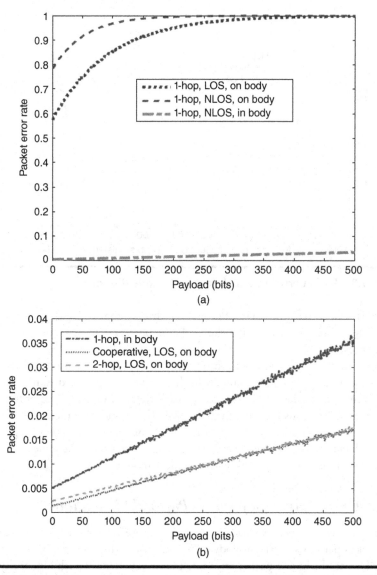

Figure 7.2 (a) Packet error rate with respect to payload, for three different scenarios: one-hop, cooperative, and two-hop scenarios with S-D separation is 100 cm, S-R separation is 50cm, and R-D separation is 70 cm. (b) Same simulation for the three different scenarios but with a LOS in communication.

communication. From the figure, cooperative communication has the smallest packet error rate due to using one relay node that increases the reliability of the system by increasing the successful transmissions. Although the cooperative scenario has a better power efficiency than the other two alternatives, in practical

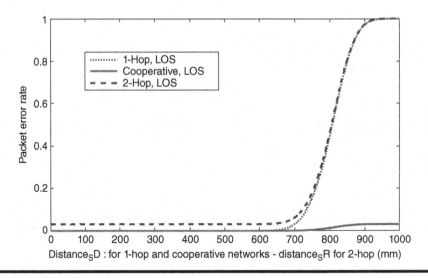

Figure 7.3 Packet error rate with respect to node distance.

payloads, which are more than 200 bits, the three scenarios have the same packet error rate. Nevertheless, the two-hop scheme performs better under longer distance thresholds.

Performance of the wireless communication in WBASNs also depends on the distance between the nodes. Namely, for higher than a specific distance threshold (~680 mm), the packet error rate cannot reach the zero value no matter how small the packet size is. This is shown in Figure 7.3. The cooperative method is very reliable and has the smallest packet error rate by using a relay node that keeps a fresh copy of data to transmit when necessary. Other schemes show the same packet error rate and will perform very weakly for distances more than 80 cm.

7.4.2 Energy Efficiency versus Payload Simulations

Energy efficiency versus payload is shown in Figure 7.4 for the one-hop scenario. Due to less path loss in the on-body communication, this scenario is the most efficient. Due to the higher packet error rate that is experienced with higher payloads, the optimum payload for in-body scheme is the least and the on-body NLOS curve stays in the middle. It is worth mentioning that for human body signals, payload is an important performance metric that determines delay in transmission of the vital body signals.

Figure 7.5 shows the same simulation results for the one-hop, cooperative, and two-hop schemes. Due to low PER of the cooperative and two-hop methods, they have considerably higher energy efficiency than the one-hop scheme. The two-hop communication method has higher efficiency than the cooperative method because of less energy consumption for transmission of every bit.

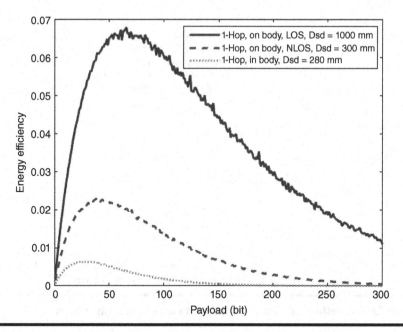

Figure 7.4   Energy efficiency of one-hop with respect to payload.

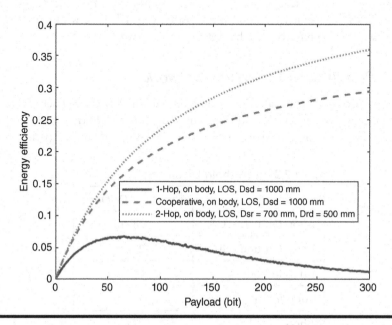

Figure 7.5   Energy efficiency comparison with respect to payload.

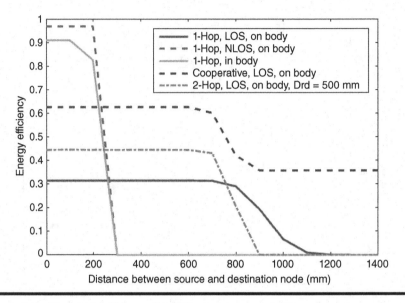

Figure 7.6 Threshold distance for different communication scenarios

7.4.3 Energy Efficiency with Respect to Distance

Figure 7.6 shows the energy efficiency of all five scenarios. From this figure, threshold distances for each scheme are extracted and summarized in Table 7.2. For large distances, the cooperative method proves to be very efficient, while for medium range, two-hop is preferred for a battery-powered network.

7.4.4 Results for a Batteryless Network

Table 7.3 lists some commonly used body signals in WBASNs. Each signal has different frequency and thus a different sampling rate. In addition, each sample is considered to be 12 bits. The harvestable energy is the amount of energy that can be

Table 7.2 Threshold Distances

Scenario	Threshold Distance
1-hop On-body LOS	78 cm
1-hop On-body NLOS	22 cm
1-hop In-body	11 cm
Cooperative	>140 cm
2-hop	82 cm

Table 7.3   Specification of Different Sensor Nodes in a WBASN

Physiological Signal	Parameter range	Maximum frequency (Hz)	Sample Interval time (sec)	Payload Sample (bit)	Harvestable Energy Per payload Transmission
Blood Flow	1–300 ml/s	20	0.025	12	10uW
ECG Signal	0.5–4 mV	250	0.002	12	0.8uW
Respiratory rate	2–50 breathes/ min	10	0.05	12	20uW
Blood Pressure	10–400 mmHg	50	0.01	12	4uW
Blood pH	6.8–7.8 pH units	2	0.25	12	100uW
Nerve Potentials	0.01–3 mV	10,000	5e-6	12	0.002uW
Body temperature	32–40 °C	0.1	5	12	2mW

harvested from ambient sources between two transmissions, while data transmission is in its inactive cycle in a duty-cycling scheme. Depending on the power of the energy harvester and the sampling interval, a specific amount of energy can be gathered in each node. Taking advantage of the network heterogeneity and assuming variable packet sizes, the scavenged energy needs to be balanced in each node. Also we conclude that the harvesting time for each node and the amount of energy available for each transmission is determined by the sampling rate of the node. For nodes with higher sampling rate, lower energy can be harvested. Therefore, higher number of samples should be aggregated in the node in order to provide more time to harvest enough energy for the next transmission. In this section, a fully integrated thermoelectric energy harvester designed for wearable devices is assumed. It has been shown that it has the highest output power among the fully integrated solutions in the literature.

Table 7.4 reflects the fact that for high-frequency signals, higher packet sizes will be feasible to send and optimal packet sizes maximizing energy efficiency will not be optimal for a batteryless WBASN. The optimal scenario with a one-hop transmission that has the fastest response with the smallest packet size is obviously LOS, because of the minimum energy consumption. In addition, the cooperative method is faster than the two-hop method.

Table 7.4 Results for the Batteryless WBASN

Network	Number of Excessive Samples in a packet, in a batteryless WBASN						
	Blood Flow	ECG Signal	Respiratory Rate	Blood Pressure	Blood pH	Nerve Potentials	Body Temp
1-hop LOS	0	35	0	3	0	13,000	0
1-hop NLOS	0	635	0	0	0	370,000	0
1-hop Implant	0	75	0	0	0	78,000	0
Cooperative LOS	0	85	0	0	0	95,000	0
2-hop LOS	0	95	0	0	0	100,000	0

7.5 Conclusion

A novel algorithm for packet-size optimization of batteryless wireless body area sensor networks is proposed. This method relies on the energy balance of each node. Models describing energy consumption at each node type, as well as, their energy harvesting models have been derived. System models and energy efficiency equations for one-hop, cooperative, and two-hop transmission schemes in a WBASN are also analyzed. Numerical results show that a one-hop transmission scenario has higher energy efficiency only for signals that have relatively large sampling times, such as blood pressure. For small sampling interval times, such as ECG signals, cooperative networks offer better energy efficiency and thus faster response. These findings, together with implementing the proposed adaptive algorithm in the topology of the network, can target topology and adaptive optimization of WBASN with various sensor types such as image and video sensors.

References

1. Demir S., Al-Turjman F. 2018. Energy scavenging methods for WBAN applications: A review. *IEEE Sensors Journal*, 8(16), 6477–6488.
2. Dong Z., Gu H., Wan Y., Zhuang W., Rojas-Cessa R., Rabin E. 2015. Wireless body area sensor network for posture and gait monitoring of individuals with Parkinson's disease. *ICNSC 2015 - 2015 IEEE 12th Int. Conf. Networking, Sens. Control*, 81–86.
3. Gui-ling G., Jia-long Y., Ying Z., Wei-xiang L. 2011. Design and implementation of sensor nodes for a wireless body area network. *2011 4th Int. Conf. Biomed. Eng. Informatics*, 3, 1403–1406.

4. Khan J.Y., Yuce M.R. 2010. Wireless body area network (WBAN) for medical applications, in *New developments in biomedical engineering*, D. Campolo, ed. Rijeka, Croatia: IntechOpen.

5. Wu T., Wu F., Redoute J.-M., Yuce M.R. 2017. An autonomous wireless body area network implementation towards IoT connected healthcare applications. *IEEE Access*, 5, 11413–11422.

6. Hamid R., Yuce M.R. 2017. A wearable energy harvester unit using piezoelectric–electromagnetic hybrid technique. *Sensors Actuators, A Phys.*, 257, 198–207.

7. Al-Turjman F., Mostarda L., Ever E., Darwish A., Shekh Khalil N. 2019. Network experience scheduling and routing approach for big data transmission in the Internet of Things. *IEEE Access*, 7(1), 14501–14512.

8. Al-Turjman F., Imran M., Vasilakos A. 2017. Value-based caching in information-centric wireless body area networks. *Sensors Journal*, 17(1), 1–19.

9. Pandey B., Jain A., Azeem M.F. 2017. Self-sustaining WBAN implants for biomedical applications. *Proc. 2016 2nd Int. Conf. Appl. Theor. Comput. Commun. Technol.* iCATccT 2016, 494–503.

10. Kappel R., Pachler W., Auer M., Pribyl W., Hofer G., Holweg G. 2013. Using thermoelectric energy harvesting to power a self-sustaining temperature sensor in body area networks. *Proc. IEEE Int. Conf. Ind. Technol.*, 787–792.

11. Domingo M.C. 2011. Packet size optimization for improving the energy efficiency in body sensor networks. *ETRI J.*, 33(3), 299–309.

12. Al-Turjman F., Killic K. 2019. LaGOON: A simple energy-aware routing protocol for wireless nano sensor networks. *IET Wireless Sensor Systems*. DOI: 10.1049/iet-wss.2018.5079.

13. Promwongsa N., Sanguankotchakorn T. 2016. Packet size optimization for energy-efficient 2-hop in multipath fading for WBAN. *Proc. Asia-Pacific Conf. Commun.* APCC 2016, 445–450.

14. Hiep P.T., Hoang N.H., Kohno R. 2015. Performance analysis of multiple-hop wireless body area network. *J. Commun. Networks*, 17(4), 419–427.

15. Khan J.Y., Mehmet R., Yuce M.R. 2010. Wireless body area network (WBAN) for medical applications. *New Developments in Biomedical Engineering*. InTech.

Chapter 8

A Cognitive Routing Protocol for WBAN

Fadi Al-Turjman

Department of Computer Engineering, Antalya Bilim University, Antalya, Turkey

Contents

8.1 Introduction 130
8.2 Related Work 131
8.3 System Models 133
    8.3.1 IoNT Architecture 133
    8.3.2 Lifetime in IoNT 135
        8.3.2.1 Lifetime Based on the Number of Alive Nanonodes 135
        8.3.2.2 Lifetime Based on the Nanonode Coverage 135
        8.3.2.3 Lifetime Based on Coverage and Alive Nanonodes 135
    8.3.3 Energy Conservation and Dead Node Issue 136
    8.3.4 Communication Model 137
        8.3.4.1 Number of Hops without Delay Constraints 138
        8.3.4.2 Number of Hops with Delay Constraints 139
8.4 Enhanced Energy-Efficient Approach (E^3A) 139
    8.4.1 Learning 140
    8.4.2 Reasoning 140
8.5 Performance Evaluation 140
    8.5.1 Shortest Path Algorithm (SPA) 141
    8.5.2 Nearest Neighbor Algorithm (NNA) 141
    8.5.3 Experimental Setup 141

     8.5.4 Performance Metrics and Parameters...142

     8.5.5 Simulation Results...143

8.6 Conclusions..147

References 147

8.1 Introduction

The wireless body area network (WBAN) is a network that provides continuous monitoring over or inside the human body for a long period and can support transmission of real-time traffic, such as data, to observe the status of vital organ functionalities [1]. This technology has found great interest in health and infrastructure monitoring; and investigations in it are still ongoing. Moreover, WBAN inception has provided enhanced and efficient solutions to various applications in biomedicine, industry, agriculture, and military applications that rely on nanotechnology science.

The field of Internet of Things (IoT) has been continuously growing, especially in the past decade. Furthermore, many technological advances have been reached in the field of nanotechnology. Combining both these fields (Internet of Things and nanotechnology), a new field termed as the Internet of Nano-Things (IoNT) has emerged. Intelligent, energy-efficient, and trustworthy wearable nanodevices can dramatically enhance and transform the human experience and interaction and perceive the world around us. However, there are certain difficulties, both conceptual and technical, that need to be traversed before this IoNT paradigm can be realized in our daily lives.

One of the main areas that this technology has been used in is the area of healthcare. Where systems for monitoring the internal well-being of the human body have been developed, these systems usually employ a vast number of nanosensors embedded in the human body which continuously communicate with each other and with the outside environment, forming the WBAN. For example, nanosensors which can monitor the glucose level of the blood have been developed for the protection of diabetes patients or possible diabetes patients [2]. Furthermore, magnetic nanosensors for the detection and profiling of erythrocyte-derived microvesicles have also been used [4]. The successful implementation of such a technology will make the monitoring of the individual health much easier by offering an all-time low-cost monitoring system.

Several IoNTs' design aspects, which stem from their unique features in terms of limited-energy constraints, short communication range, and low processing power, needed to be incorporated into their routing protocols in order to realize the IoNT paradigm. Different challenges against routing protocol design in terms of energy are still being investigated with no currently fully developed solutions. Nanonetworks consume energy in almost all processes. They consume energy while making data transmission, data sensing, and data processing. There have been a few attempts towards achieving energy efficiency in such networks via wireless multi-hop networking [2–5]. However, such schemes either assume static network topology, which renders these schemes impractical for real-life network implementation, because nanonetworks exhibit random topology due to the mobility of nodes, or are restricted to two-hop from source to the sink routing schemes.

Due to the fact that nanonetworks' sensors are usually limited in their processing power, communication range, and energy capabilities, design and implementation of routing algorithms are considered a nontrivial task. In this chapter, we propose an enhanced energy-efficient algorithm (E^3A). E^3A is a cognitive data delivery approach that addresses the challenges of data delivery in IoNT networks. The proposed approach caters to the grid-based distribution of the employed nanosensors on the monitored object/body organ to efficiently and effectively cope with the dynamicity of nanonetwork topology [6].

The rest of the chapter is organized as follows. Section 8.2 reviews previous related studies. Section 8.3 discusses our system models. Section 8.4 describes our proposed routing approach for IoNT paradigm. Section 8.5 provides performance evaluation for the proposed approach. Finally, Section 8.6 provides the conclusions and future directions.

8.2 Related Work

Connectivity in IoNT involves finding reliable routes from the event sensory node to the IoNT gateway. Utilizing duty-cycling, a routing algorithm can be designed to significantly balance the network load and optimize the energy consumption, especially in energy-harvesting IoNT networks. At the same time, the viability of duty cycling allows the network to circumvent elements and zones rendered temporarily unavailable, whether due to energy unavailability, medium variations, or mobility and body changes. In IoNT networks, a critical problem to overcome is the uneven energy consumption across the network where elements near the gateway would deplete their energy faster than those at a distance. And thus, the routing protocol should be coupled with MAC layer protocol through cross-layer design. Feedback from the MAC and physical layers, in addition to the residual energy and current load of the IoNT nodes, shall be utilized to identify and bypass critical links. Reducing the ratio of lost packets during channel impairments is an important reliability objective as well. The MAC layer is usually expected to adapt the number of retransmissions depending on channel quality. Current medium access control (MAC) protocols, however, typically limit the number of backoffs and retransmissions. Due to the peculiarities of the terahertz (THz) communication, which enables enormous bandwidth and negligible transmission delay, the probability of collision is considerably less in IoNT than that of traditional wireless networks. Moreover, the unique characteristics of THz communication band and specific challenges of IoNT, such as energy constraints and random network topology, prevent the direct implementation of traditional wireless sensor networks' (WSNs) routing schemes. In this section, we discuss traditional sensor network protocols with preferable features to be considered in any IoNT routing protocol in order to reduce energy consumption.

Multipath versus single path in traditional WSNs: Recent studies show that multipath routing protocols are better than single-path routing protocols in terms of quality of service (QoS) parameters. Multipath routing protocols provide higher probability to achieve packet reachability while utilizing redundant paths to the sink. For example, sequential assignment routing (SAR) was the first routing protocol developed for WSNs [30] in which QoS issues were considered based on three factors: energy conservation, QoS parameters, and level of packet-priority traffic flow. Nevertheless, the disadvantage of SAR is that the creation mechanism of the multipath causes additional node energy depletion. Thus, it is not suitable for energy constrained IoNT. The reliable information forwarding using multiple paths (ReInForM) protocol as described in [7] employs probabilistic flooding to deliver information awareness packets and service at desired priority levels of reliability, at a proportional cost for sensor networks. Unfortunately, this protocol does not consider the delay deadlines of the packet when selecting the multiple paths. And thus, it duplicates the packet retransmission, which causes a high cost-of-energy consumption and the occupation of useful channel bandwidth. Meanwhile, the geographic adaptive fidelity (GAF) protocol makes it possible to optimize the performance of sensor networks by specifying the equivalent nodes according to a set of forwarding packets [8]. However, it is crucial in GAF to precisely identify the sensors' geo-positions, which is a challenge in IoNT. On the other hand, the geographic and energy-aware (GEAR) routing protocol is used for improving the efficiency of the energy by forwarding queries to specific aimed regions [9]. In this solution, the sensors must have also localization hardware such as a GPS unit. However, these localization units are usually inaccurate and lead to dramatic degradation in energy consumption. In addition, nanonetworks cannot utilize traditional localization techniques; i.e., new proposals for localization of nanonodes is still an open area of research. In the nearest neighbor algorithm (NNA), proposed in [10], when a packet is transmitted from one node to another, it follows the shortest path. NNA assumes that if a packet always follows shorter path, it will use shortest path until it reaches destination node. This algorithm uses four-direction transmission only. So, it actually does not consider shortest path, it considers shortest neighbor relay node in order to send data. As a result of this, hop count unnecessarily increases and also energy consumption of the network nodes is negatively affected by increasing the hop count. Meanwhile, in the shortest path algorithm (SPA), discussed in [9], when a packet is transmitted from a node, the algorithm calculates the shortest path from recent node to destination instead of node to node, and the packet follows this path until it reaches the destination. In other words, SPA considers shortest path to destination rather than shortest neighbor of the relay node. And thus, the hop count decreases and energy consumption of the nodes decreases in comparison with NNA. Moreover, radio interference, antenna shape/orientation, and environmental factors in IoNT may vary during the network lifetime and affect link quality between the nanosensor nodes. These variations result in asymmetric links between the nanosensor nodes [12][13]. Therefore, these routing approaches shall be adapted to estimate the link quality as well while finding their optimal paths in nanonetworks.

In this chapter, we proposed an enhanced energy-efficient algorithm (E^3A) as a routing protocol. It considers the remaining energy at the routing nodes (RNs). If one of these neighboring RN's energy is below half of the initial energy, a new path will be determined and new packets will follow this new path.

8.3 System Models

In this section we list the assumed system models for the proposed E^3A approach toward prolonged lifetime in IoNT.

8.3.1 IoNT Architecture

The typical communication range in the IoNT nanonetworks is expected to be between 10 nm and 100 nm for terahertz-based communication [18]. This means that the transmission range is extremely limited, which makes multi-hop routing a critical aspect for nanonetworks. Furthermore, the direction of a communication route is not deterministic and is dependent on the drift velocity of the nanonode, which may lead to unnecessary communication delays. This necessitates effective nanonetwork architectures. In this chapter we propose a hierarchical nanonetwork architecture based on [19] for bio-inspired applications consisting of the following components:

- Nanonodes: These are the smallest and simplest nanodevices. Due to the limited energy, limited memory, and reduced communication capabilities, they can only perform simple computation tasks and can transmit over very short distances. These nodes are usually composed of tiny sensor and communication units.
- Nanorouters (NR): These are the nanodevices with slightly larger computational resources than nanonodes and can aggregate information from limited nanomachines and also can control the behavior of nanonodes by sending extremely simple orders (such as on/off, sleep, read value, etc.). However, this would increase their size; thus, their deployment would be more invasive.
- Nano-micro interface: These are used to aggregate the information forwarded by nanorouters and send the information to the microscale devices. At the same time, they can send the information from microscale to nanoscale. Nano-micro interfaces are hybrid devices not only able to communicate in the

nanoscale using the nanocommunication techniques, but also can use classical communication paradigms in micro/macro communication networks. Hence, in this chapter we define these nodes as cognitive relay nodes (CRNs).

■ Gateway (GW): It enables the users to control/monitor the entire system remotely over the Internet. It is a hybrid device not only able to communicate in the nanoscale, but also can use classical communication paradigms in micro/macro communication networks. It is also known as the global cognitive nodes (GCN).

We define elements of cognition to implement the functionality of the observe-analyze-decide-act (OADA) loop [11]. Reasoning and learning constitute the elements of cognition, which, when implemented in specialized nodes of the network, can help in making cognitive decisions and making the nanosensor network a cognitive one. In the proposed IoNT paradigm, CRNs and GCNs are the specialized nodes that implement the elements of cognition. On the other hand, nanosensor nodes only host raw sensed data. Raw sensed data is represented and stored in attribute-value pairs. This representation facilitates named-data identification to locate the user-requested information. Thus, the two main simple tasks of the nanosensor nodes are: (i) sensing raw data, and (ii) storing/representing sensed information in frame-based knowledge representation, where a frame is defined as a hierarchical data-structure with inheritance [11]. It has slots which are function-specific cells for data. In nanosensor nodes, these function-specific cells store sensor attribute-value pairs. In CRNs, they store more information, such as the one-hop neighbor CRNs and the associated values of QoI attributes in the last communication round. Information accumulated over several rounds of information transmission leads to the formation of a knowledge base (KB), which can be looked up by the reasoning mechanism to make quick decisions on choosing the data delivery path which satisfies the QoI delivered to the end user. Nanosensors communicate with nanorouters and cognitive relay nodes (CRNs). RNs communicate with nanosensors and CRNs to act as intermediate nodes that gather information from nanosensors and forward it to their associated CRN neighbors. CRNs perform two main functions: (i) gathering sensory data from nanosensor nodes and forwarded information from RNs, and (ii) delivering data based on required QoI levels/attributes. CRNs also function as caches to store the data as it travels through the network. CRNs make use of the nanosensor attributes to identify the relevant data, similar to the named data-object search in information-centric networks [26]. The required QoI attributes are based on the type of traffic flow generated as a result of the end user's request. As for dealing with the QoI attribute requirements, an analytic hierarchy process (AHP) [11] is implemented as the reasoning element of cognition to make the cognitive decision in the CRNs. They deliver data over multi-hop paths to the GCN. GCNs in this architecture have the following main functions: They receive user requests and synthesize it to identify the following information: application traffic type, requested nanosensor attributes, and QoI attribute priorities. They broadcast the synthesized information to the CRNs so that they may process it further to gather the requested information from the network. Once the network returns the requested information, GCNs process it to determine if the QoI provided by the network meets with the user requirements, and deliver information with acceptable QoI to the end user. They also determine when the network is no longer able to deliver useful information from the network, thus flagging the end of life of the network. Consequently, we delegate the heavy cognition tasks to more capable and specialized nodes (e.g. the CRNs and GCNs) without affecting the original nanodevices' simplicity.

8.3.2 Lifetime in IoNT

We can evaluate the nanonetwork lifetime in the IoNT paradigm by three models, as listed below.

8.3.2.1 Lifetime Based on the Number of Alive Nanonodes

Several variants do exist with this model. The simple model identifies the time until the death of the first nanonode in the network as the lifetime of the network. Another variant evaluates lifetime until the death of 'k' out of 'n' nanonodes in the network, where $k < n$. Thus, the nanonetwork lifetime is the range between the death of 'k' nodes out of 'n' non-critical nodes [20]. It's worth pointing out also that the utilized nanonodes in IoNT are supposed to operate based on an energy-harvesting mechanism, and thus, a nanosensor node will be considered dead only and only if this mechanism stops functioning or physical damage was experienced. However, it should be remarked also that while a nanosensor is harvesting energy, it cannot perform any communication task. This makes every nanosensor able to transmit only once per round. The lifetime of a nanonetwork will depend on energy harvesting/depletion rate β_{harv} [12]. This rate can affect the probability of having a relay node in the neighborhood.

8.3.2.2 Lifetime Based on the Nanonode Coverage

This model defines the lifetime of the network in terms of the coverage of region of interest. If it is used to ensure that all points inside a region of interest are covered, it is denoted by volume coverage. When an identified number of target points are to be covered, it is denoted as target coverage. Normally, a region of interest should be covered by at least one nanonodes cluster.

8.3.2.3 Lifetime Based on Coverage and Alive Nanonodes

This type of metric is mostly found in ad-hoc nanonetworks. In this option, lifetime is defined as the period during which most of the nodes are connected with each other. Because in IoNT each node has to communicate with a gateway node, this metric cannot be used as is. Another issue with this metric is that the lifetime is based on the total number of packets transmitted to the gateway. Nevertheless, in most of the related works this metric becomes useless [21].

8.3.3 Energy Conservation and Dead Node Issue

Energy conservation is one of the most important issues in IoNT design. Nanonodes are restricted in carrying out the network layer functions; their main task is to flood the data to their one-hop routers. Hence, the multi-hop forwarding between source and gateway is normally performed by RNs, which have relatively higher capabilities than nanonodes. In the following, we characterize the energy cost of a RN:

$$E_{RN} = C(T * (E_{TX}) + R * (E_{RX})) \tag{8.1}$$

Most of the energy consumption at the RN is due to data communication, indicated by E_{TX} for energy consumed during transmission and E_{RX} for energy consumed during data reception. C represents cost function of the energy consumed, T represents the number of transmitted packets, and R represents number of received packets. As we mentioned before, the CRN's main function is data aggregation and routing the received traffic from RNs. The capabilities of the CRN are higher than that of RN, and hence, it is expected to consume additional energy compared to regular RNs. In fact, the additional energy consumption can be due to two factors: one is the routing protocol overhead incurred while making a cognitive decision based on the nanonetwork feedback during the learning process, and the other one is due to the increased transmission power of the CRN for extended coverage purposes as follows:

$$E_{CRN} = C(T * (E_{TX}) + R * (E_{RX})) + C(Ag * (E_{ag}) + C(P * (E_{cog} - E_{pro})) \tag{8.2}$$

In Eq. (8.2), T, R, Ag, and P represent the total number of packets that are transmitted, received, aggregated and processed by the cognitive elements respectively, in each transmission round. $C(T * (E_{TX}) + R * (E_{RX}))$ is the energy cost incurred during data transmission and reception, $C(Ag * (E_{ag}))$ represents the energy cost incurred during data aggregation and $C(P * (E_{cog} - E_{pro}))$ indicates the energy cost due to protocol and processing overhead during the cognitive processes. Forming Eq. (8.2) in terms of the energy cost of RNs we get:

$$E_{CRN} \geq E_{RN} + C(Ag * (E_{ag}) + C(E_{cog} - E_{pro}) \tag{8.3}$$

If the RN and CRNs use the same power amount to transmit a data packet, the equality sign becomes positive in Eq. (8.3). In order to ensure that the energy cost of *CRNs* does not stabilize the advantages it offers, based on adapting to the dynamic traffic flows and network topology alteration, the cost can be optimized by maximizing the number of RNs and minimizing the *CRNs* in the deployment [14][15].

In this work, we refer to the one-hop neighbors' communication as the first tier of nodes. Since no other node can reach the monitoring station directly, traffic from every other node will have to be forwarded in the last hop by one of these first-tier nodes. Similarly, the two-hop neighbors of the monitoring station will forward data for all nodes except the one-hop neighbors and themselves [17]. If the spatial distribution of nodes is assumed to be uniform, then the traffic load is spatially and equally distributed. Each first-tier node will forward hardly the same amount of traffic, and all first-tier nodes will die at times very close to each other, after the network is first put into operation. Once all of the first-tier nodes are dead, no other node will be able to send data to the gateway node, and the lifetime of the network will be over. Increasing the number of nodes in the network accentuates this effect, since there is more traffic to forward and the first tier of nodes has a smaller share of the total energy budget.

In general, the network death in IoNT can be associated with several cutoff criteria, such as the first node death, the percentage of dead nodes, or the number of dead nodes rising above a specific level where the routing to the sink node is no longer possible [22]. Nevertheless, as we are experimenting with the clustering-based protocols [23], in which the energy is evenly distributed throughout the mobile nanonetwork, we consider the first scenario for the definition of the network lifetime. Because when the first node dies, the number of dead nodes increases in the later rounds, and within 5–10 rounds the whole network becomes nonoperational. According to preliminary results, non-position-based routing protocols outperform geo-based protocols in terms of network lifetime. The primary reason for this behavior is that location-based protocols consume energy in terms of localization services. Moreover, the number of control messages plays a vital role in the network lifetime.

8.3.4 Communication Model

Considering that the communication is at nanoscale, the study of the communication in very short range is essential [22]. And hence, we consider the proposed path loss formula in [22] at THz, which has two parts: the absorption path loss and the spread path loss. Four different power spectral densities (p.s.d.) were studied by authors in [22] i.e., optimal p.s.d., flat p.s.d., the Gaussian pulse, and the p.s.d. for the case of the transmission window at 350 GHz, which concluded that for the very short communication range in bio-inspired applications, very high transmission bit-rates can be supported, up to terabits per second, indicating the promising future of nanocommunication.

Meanwhile, energy-aware IoNT frameworks depend heavily on two main principles in their communication design. The first principle is the number of hops without the delay constraint. The second, is the number of hops with the delay

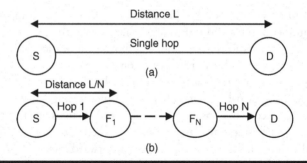

Figure 8.1   Single-hop and multi-hop communication systems [16].

constraint. Figure 8.1 shows single-hop communication system in (a) and multi-hop communication system in (b).

8.3.4.1 Number of Hops without Delay Constraints

If there is no delay constraint on the system, the highest achievable transmission rate is given by Eq. (18) in [22].

$$C_{ref}(d) = \sum_{i}^{n} \Delta f \, log_2 \left[1 + \frac{S(f_i)A(f_i,d)^{-1}}{N_0(f_i,d)} \right] \tag{8.4}$$

The bandwidth is divided into i sub-bands; the i-th sub-band is centered around frequency $f_i, i = 1, 2, \dots$ and it has width Δf. If the sub-band width is small enough, the channel appears as frequency-nonselective and the noise p.s.d. can be considered locally flat. The resulting capacity in bits/s is then given by $C_{ref}(d)$ in Eq. (8.4), where d is the total path length, S is the transmitted signal p.s.d., A is the channel path loss, and N_0 is the noise p.s.d. Based on [24], we find the end-to-end capacity of N hops path by

$$C_{e2e} = C_I (1 - F_{AVG})^N \tag{8.5}$$

where C_I is the channel capacity contributed by the first hop, N is the hop count determined by the forwarding scheme, and F_{AVG} is the average capacity loss factor per hop. The value of F_{AVG} is calculated as follows:

$$F_{AVG} = F\left(\frac{d_0 - d_{AVG}}{d_0} \right) \tag{8.6}$$

where F is the capacity loss factor and d_0 is a constant that denotes the reference distance from source-to-sink.

8.3.4.2 Number of Hops with Delay Constraints

The predictions about the preferred number of hops made in the previous section were based on the assumption that the block lengths used by channel codes can be randomly large [25]. In many applications there is a strict limit on the tolerable end-to-end latency. There are several factors of latency in nanocommunication systems. In the following we list these factors:

- Waiting for the data source to emit enough bits to form a block of a desired length;
- Processing delay caused by encoding/decoding the information bits for transmission;
- Transmission and reception of the whole encoded message.

If the communication system involves multiple hops, the latter three elements are repeated several times, increasing overall latency. To compensate for this, shorter block lengths must be used at a cost of reduced error-correcting capabilities at each link [14].

8.4 Enhanced Energy-Efficient Approach (E^3A)

In this section we propose a novel energy aware data delivery approach for the energy-constrained IoNT, namely the E^3A. Data delivery decisions in the E^3A are based on observing the dynamically changing topology of the network. Since the learning process might be too slow to respond/converge before further changes take place in the network, we choose a heuristic search strategy, called AHP, to aid in making quicker decisions that suit the randomly deployed nanosensors. With the help of a model of the original network topology combined with the currently observed changes, we make use of the AHP heuristic algorithm [11] to identify the nodes that can be used for data delivery to the sink (GCN). This approach is useful when a problem is to be solved repeatedly with the same goal at GCN, but with different initial states at CRNs. In the AHP algorithm, RNs and CRNs are initially assigned heuristic values at the time of deployment based on their proximity to the sink. Nodes that lie at one-hop distance to the sink have the highest probability of successful data delivery to the sink. Hence, they are given the highest heuristic weight (0.1 in our study). Nodes lying further away from the sink are given lower weights (0.05), so that they have lesser influence on the heuristic decision making. However, they are not assigned a zero weight, because these nodes will still be able to participate in multi-hop routing in case the nodes with direct access to the sink become unavailable due to poor link conditions, network congestion, or node deaths. The higher the weight of the heuristic, the higher will be the chance that the node will be chosen for data transmission to the sink. These values remain fixed at each RN and CRN until the nodes die, at which time the heuristic values are made "0" as they do not influence the heuristic decision anymore.

8.4.1 Learning

Learning is used in our E^3A approach in order to determine the most appropriate paths towards the GCN that satisfy the nanonetwork requirements. This cognition element uses a direction-based heuristic to determine the data delivery path through RNs that lies in the direction of the GCN. Hence, each time a CRN has to choose the next hop, the direction-based heuristic eliminates RNs that increase the distance between the current RN and GCN. Knowledge of the positions of the CRN and its one-hop RNs is used by the heuristic to determine the set of such RNs, which we call forward-hop RNs. Thus, the forward-hop RNs of a CRN identified by the direction heuristic is constituted by those RNs that reduce the distance between the CRN and the GCN. This information is stored in the CRN for use in the next transmission rounds. Thus, the direction-based heuristic, along with feedback from the network about the chosen paths, helps the CRNs to learn data delivery paths to the sink as the network topology changes.

8.4.2 Reasoning

In the E^3A approach, we assume a modified version of the analytic hierarchy process (AHP) [26] for implementing the reasoning element of cognition in the IoNT. AHP supports multiple-criteria decision making while choosing the data path. For example, if we have delay-sensitive data, the node which provides the lowest latency will be chosen even though it might degrade other metrics such as the network energy or throughput. If two next-hops guarantee the same latency, then the next attribute to compare will be energy, and then throughput, assuming that energy is the next desired attribute in the nanonetwork. AHP provides a method for pairwise comparison of each of the attributes and helps to choose the node that can provide the best network performance in the long run. The following subsequent example has more details on the utilized AHP. While AHP calculations help in deciding the next-hop, it also helps in planning for future actions. The CRNs are able to store the calculated values of the nanonetwork attributes, which can be used in future transmission rounds. Hence, these values are not necessarily calculated at every transmission round.

8.5 Performance Evaluation

In this section, we evaluate the performance of the proposed E^3A. We use SPA and NNA algorithms as baseline evaluation algorithms. Based on the aforementioned system models, we summarize these two baselines' categories as follows.

8.5.1 Shortest Path Algorithm (SPA)

The first baseline category in this research is the SPA. As we mentioned before, it is one of the most well-known works in routing. It is used in two cases:

Case 1, polling, when GCN requesting a packet from sensor nodes. In this case, the GCN randomly chooses one RN and adds the index of RN to the request package. If the range of GCN covers the selected RN, it sends the request packet directly to this RN. If not, it sends the request packet to the RN that is nearest to the target RN. Then from that RN, the packet is transmitted to the RN that was selected originally.

Case 2, pushing, occurs when the data packet is transferred from nanosensors to the GCN. The aim is to transmit the data packet to the nearest RN that is in the range of the GCN. With this routing approach, the shortest path is calculated and after that, data packet is hopped from the current RN to the next one. When the packet reaches to the RNs near GCN these RNs transmit the packet back to the GCN.

8.5.2 Nearest Neighbor Algorithm (NNA)

The NNA approach is one of the previous works which has been designed mainly for wireless sensor networks and also is an efficient version of the shortest path algorithm. This approach calculates the shortest path but the transmission has to be occurred only towards up, down, left or right directions. In short, there is no diagonal moving, so this surely increases the hop counts, and this justifies the increment in energy consumption.

In the aforementioned two baselines, SPA represents a straightforward approach in cutting down unnecessary energy consumption while choosing the shortest path. On the other hand, NNA chooses one of the RNs that has a non-diagonal connection. Using NNA increases hop count, so energy consumption is increased and network life time is decreased. The neighbor RNs' energy level and distance from GCN are compared in E^3A. It chooses RN, which meets the requirements for transmission and maintains the most energy-efficient topology by applying the AHP algorithm for the prolonged network lifetime. It is typical to have a deterministic placement for the RNs in bio-inspired applications where a specific area of the skin is targeted, for example, and these nanodevices are planted [29]. However, the selection of which RN to route the sensed data through varies based on the utilized routing algorithm. Thus, we compare our proposed E^3A routing algorithm with both NNA and SPA in this research. A detailed description of our experimental setup is given in the following section.

8.5.3 Experimental Setup

In order to limit our search space, we assume a virtual grid, where SNs are placed on the grid vertices. We assume up to 1500 total SNs communicate with one GCN

Table 8.1   Simulation Parameters and Values [25, 29]

Parameter	Value
Target area	10 mm × 10 mm
Number of nodes	SNs: 100, RNs: 16, GCN: 1
Communication range	SN: *142 nm*, RN: *300 nm*, GCN: *500 nm*
Initial energy	SN: *31 pJ*, RN: *110 pJ*, GCN: *Unlimited*
Energy consumption	SN and RN (Receiving): *31.2 pJ/bit* SN and RN (Transmitting): *53.8 pJ/bit*

via 36 RNs. We used NS3 as a simulation tool for this purpose. The simulation is processed in three platforms, which are Windows, Linux, and OSX for validation purposes. We executed our simulation 100 times for each experiment and plotted the average results. More details about our assumed simulation parameters are summarized in Table 8.1. It's worth pointing out here that one of the most attractive applications of nanonetworks is in bio-applications, due to its advantages of size, bio-compatibility and bio-stability. Nanodevices spreading over the human body can monitor human physical movement. For example, nanopressure sensors distributed in the human eyes can detect intra-ocular pressure for the early diagnosis and treatment of glaucoma to prevent vision loss [27]. At the same time, the nanodevices deployed in the bones can monitor the bone growth in young diabetes patients to protect them from osteoporosis [27]. Furthermore, nanorobots inside the biological tissues can detect and then eliminate malicious agents or cells, such as viruses or cancer cells, hence making treatment less invasive in real time [28]. Moreover, networked nanodevices will be used for organ, nervous track, or tissue replacements.

8.5.4 Performance Metrics and Parameters

In this research, we assess our proposed E^3A in terms of the following main metrics:

- **Network Lifetime (msec)** is number of transmission rounds until all closest RNs to GCN are dead, based on the above-discussed energy harvesting/depletion rate β_{harv}.
- **Remaining energy (pJ)** is energy level at the end of network lifetime in the RNs and SNs.
- **Request time (msec)** is the time that the request is made periodically.
- **Success rate** is number of successfully responded requests over total number of transmission rounds during the network's lifetime.
- **Failure rate** is number of failed transmissions to GCN over total number of transmission rounds during the network's lifetime.

■ *Delay time* is the delay (latency) from event occurrence until a desired outcome is achieved, such as storing a temperature value on the local GCN memory after being retrieved from the corresponding data source.

Also, we assess our proposed E^3A in terms of two main parameters:

■ *Total number of transmission rounds* is total number of transmission rounds between nodes during the network lifetime, where in every round at least one nanosensor is reporting to the GCN.
■ *GCN request time (seconds)* represents time to receive a response from the RN.

8.5.5 Simulation Results

In this study, we are interested in examining the nanonetwork performance when the size of the nanofunctional devices goes down to milli/nanoscale. Accordingly, a tissue cube is assumed for the considered bio-applications, since the tissue size (10 mm × 10 mm × 10 mm) is comparable to THz wavelength [26]. As we mentioned before, we assume 36 RNs (0.1 mm^3 each) and 1 GCN in the area of 10 mm × 10 mm. GCN has bidirectional connection with the closest RNs to GCN, which are RN_{14}, RN_{15}, RN_{20}, and RN_{21}. Also, GCN has unidirectional connection with rest of the RNs which are in range of GCN.

Analysis based on the depicted deployment in this study shows that during the network lifetime, paths from a source RN to the GCN changes according to RNs' remaining energy on the E^3A. As a result of this, the network topology and hop count randomly change and there is no conclusive hop count for E^3A. This makes our proposed approach more adaptive to dynamic topologies in the IoNT. We remark also that hop counts are related mainly with the packet delay rate, so more hop count values mean more delay. And hence, we can find latency time for these three approaches in Figure 8.2. It is clear that E^3A has higher latency in

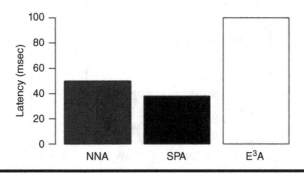

Figure 8.2   Comparison of latency.

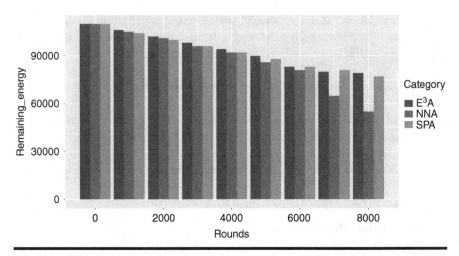

Figure 8.3   Comparison of average remaining energy level at RNs vs. total transmission rounds.

comparison to the other baselines. The reason for that is because it determines the route the packet should follow according to the energy ratios of the nearest RNs. The routing approach which has the least latency time is SPA.

On the other hand, we plot the average energy levels per round in Figure 8.3. It is clear that E^3A is more efficient than both SPA and NNA in terms of lifetime. E^3A saves more energy than SPA and NNA algorithms. In Figure 8.4, the bar chart shows the network lifetime according to the three stated algorithms. The X-axis shows algorithm type while the Y-axis shows network lifetime in seconds for the aforementioned network topology and values. While NNA and SPA have the same lifetime, E^3A has more lifetime than these two algorithms.

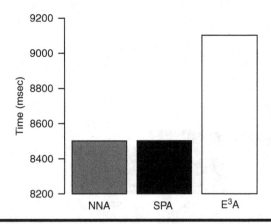

Figure 8.4   Comparison of network lifetime (msec).

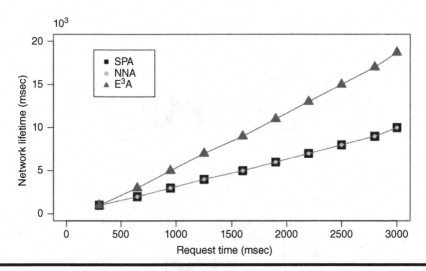

Figure 8.5   Comparison of network lifetime vs. the request time (msec).

In addition, network lifetime also depends on the GCN's request time. The line chart in Figure 8.5 shows different request times in seconds, and their effect on the network lifetime according to the different routing techniques: E^3A, SPA, and NNA. The Y-axis indicates network lifetime and the X-axis indicates request time. As the packet request intervals of GCN increase, the network lifetime normally also increases. The reason for the difference in graph is the algorithms' durability against energy spending. Although NNA and SPA have same lifetime, E^3A has extended lifetime in comparison to these two algorithms. In Figure 8.6, the Y-axis

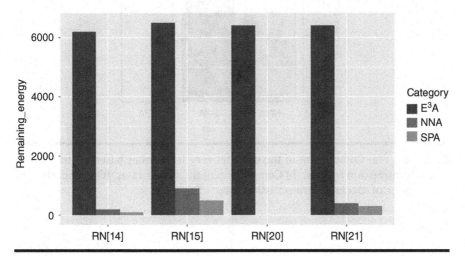

Figure 8.6   Comparison of one-hop RNs' energy level.

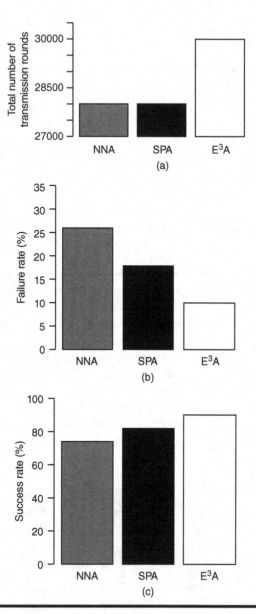

Figure 8.7 (a) Comparison of the data delivery techniques based on total number of transmission rounds. (b) Comparison of the failure rates. (c) Comparison of the number of successful transmission rate.

represents the energy level of specified RNs and the X-axis represents the specified RNs and the algorithm types. When we compare the one-hop RNs energy level with respect to these algorithms. E^3A increases the network lifetime and it is better in energy saving.

When we compare these algorithms in terms of the number of transmission rounds, it can be clearly observed from the simulation results in Figure 8.7(a) that E^3A outperforms NNA and SPA. That is because of the lower failure rate experienced while applying E^3A, as shown in Figure 8.7(b), where the number of failed transmissions from E^3A is lower than the others by at least 10%. Consequently, by looking at Figure 8.7(c), we can see that if conditions are same (e.g. same number of processed requests), our approach outperforms the other approaches with the same percentage. Since Figure 8.7(c) demonstrates the quantity of effective successful transmission rates for each routing technique, we can conclude that our E^3A approach outperforms the other two approaches, and this makes it a good candidate for nanonetworks in bio-inspired applications.

8.6 Conclusions

In this chapter, we investigated routing techniques for the IoNT paradigm in terms of energy consumption and hop counts. We proposed a novel approach for nanonetworks in IoNT, called E^3A. We found that SPA and E^3A save considerable amount of energy. We conclude that the nanonetwork lifetime has inverse proportion with the number of hop counts. Moreover, we showed how the hop counts can be used to illustrate instantaneous delays and average delays of the nanorouters. Furthermore, we showed how the E^3A algorithm provides the longest network lifetime. Both SPA and E^3A are efficient in terms of transmission and energy consumption, but the overall results show that E^3A outperformed the other two baseline algorithms.

References

1. Al-Fagih A., Al-Turjman F., Alsalih W., Hassanein H. 2013. A priced public sensing framework for heterogeneous IoT architectures. *IEEE Transactions on Emerging Topics in Computing*, 1(1), 135–147.
2. Ali S., Madani S. July 2011. Distributed efficient multi hop clustering protocol for mobile sensor networks. *International Arab Journal of Information Technology*, 8(3), 302–309.
3. Al-Turjman F., Hassanein H., Ibnkahla M. January 2015. Towards prolonged lifetime for deployed WSNs in outdoor environment monitoring. *Ad Hoc Networks*, 24(A), 172–185.

4. Hasan M., Al-Turjman F., Al-Rizzo H. 2017. Optimized multi-constrained quality-of-service multipath routing approach for multimedia sensor networks. *IEEE Sensors*, DOI: 10.1109/JSEN.2017.2665499.

5. Pierobon M., Jornet, J.M., Akkari N., Almasri S., Akyildiz I.F. 2014. A routing framework for energy harvesting wireless nanosensor networks in the Terahertz Band. *Wireless Networks*, 20(5), 1169–1183.

6. Akyildiz I.F., Brunetti F., Blázquez C. 2008. Nanonetworks: A new communication paradigm, *Comput. Netw.*, 52(12), 2260–2279.

7. Deb B., Bhatnagar S., Nath B. 2003. ReInForM: Reliable information forwarding using multiple paths in sensor networks. In *Proc. IEEE Conf. Local Computer Netw.*, 406–415.

8. Hasan M., Al-Rizzo H., Al-Turjman F. 2017. A survey on multipath routing protocols for QoS assurances in real-time multimedia wireless sensor networks. *IEEE Comm. Surveys and Tutorials*, 2017. DOI: 10.1109/COMST.2017.2661201.

9. Al-Turjman F. 2018. QoS–aware data delivery framework for safety-inspired multimedia in integrated vehicular-IoT. *Elsevier Computer Communications Journal*, 121, 33–43.

10. Samet H. February 2008. K-Nearest Neighbor Finding Using MaxNearestDist. *IEEE Transactions on Pattern Analysis and Machine Intelligence.* 30(2), 243–252.

11. Al-Turjman F. 2017. Cognition in information-centric sensor networks for IoT applications: An overview. *Annals of Telecommunications*, 72(1), 3–18.

12. Al-Turjman F. 2018. Performability in the Internet of Things. *Published with Springer*, Switzerland AG. ISBN 978-3-319-93556-0.

13. Al-Turjman F., Hassanein H., Ibnkahla M. 2015. Towards prolonged lifetime for deployed WSNs in outdoor environment monitoring. *Elsevier Ad Hoc Networks Journal*, 24(A), 172–185.

14. Singh G., Al-Turjman F. 2016. A data delivery framework for cognitive information-centric sensor networks in smart outdoor monitoring, *Computer Communications*, 74(1), 38–51.

15. Al-Turjman F., Radwan A. 2017. Data delivery in wireless multimedia sensor networks: Challenging & defying in the IoT Era. *IEEE Wireless Communications Magazine*, 24(5), 126–131.

16. Liaskos C., Tsioliaridou A. 2015. A promise of realizable, ultra-scalable communications at nano-scale: A multi-modal nano-machine architecture. *IEEE Transactions on Computers*, 64(5), 1282–1295.

17. Choudhury S., Al-Turjman F. 2018. Dominating set algorithms for wireless sensor networks survivability. *IEEE Access Journal*, 6(1), 17527–17532.

18. Agoulmine N., Kim K., Kim S., Rim T., Lee J-S., Meyyappan M. 2012. Enabling communication and cooperation in bio-nanosensor networks: Toward innovative healthcare solutions. *IEEE Wireless Communications*, 19(5), 42–51.

19. Akyildiz I.F., Jornet J.M. December 2010. The Internet of nano-things. *IEEE Wireless Commun.*, 17(6), 58–63.

20. Baranidharan B., Santhi B. November 2011. An evolutionary approach to improve the life time of the wireless sensor networks. *Journal of Theoretical and Applied Information Technology*, 33(2), 177–183.

21. Al-Turjman F. 2018. *Wireless sensor networks: Deployment strategies for outdoor monitoring*. New York: Taylor and Francis, CRC. ISBN 9780815375814.

22. Jornet J.M., Akyildiz I.F. October 2011. Channel modeling and capacity analysis for electromagnetic wireless nanonetworks in the terahertz band. *IEEE Trans. Wireless Commun.*, 10(10), 3211–3221.

23. Al-Turjman F. 2016. Hybrid approach for mobile couriers election in smart-cities. In *Proc. of the IEEE Local Computer Networks (LCN)*, Dubai, UAE, 507–510.

24. Yu H., Ng B., Seah W. 2015. Forwarding schemes for EM-based wireless nanosensor networks in the terahertz band. *ACM Proc. of the 2nd Int. Conf. on Nanoscale Computing and Communication*. Boston, MA, USA, 1-6.

25. Demir S., Al-Turjman F. 2018. Energy scavenging methods for WBAN applications: A review. *IEEE Sensors Journal*, 18(16), 6477–6488.

26. Singh G., Al-Turjman F. 2016. Learning data delivery paths in QoI-aware information-centric sensor networks. *IEEE Internet of Things Journal*, 3(4), 572–580.

27. Sitti M., Ceylan H., Hu W., Giltinan J., Turan M., Yim S., Diller E. 2015. Biomedical applications of untethered mobile milli/microrobots. *Proceedings of the IEEE. Institute of Electrical and Electronics Engineers*, 103(2), 205–224.

28. Santagati G.E., Melodia T. 2014. Opto-ultrasonic communications for wireless intra-body nanonetworks, *Nano Commun. Netw.*, 5(1), 3–14.

29. Yang K., Pellegrini A., Munoz M., Brizzi A., Alomainy A., Hao Y. 2015. Numerical analysis and characterization of THz propagation channel for body-centric nano-communications. *IEEE Trans. Terahertz Technol.*, 5(3), 419–426.

30. Sohrabi K., Gao J., Ailawadhi V., Pottie G.J. 2000. Protocols for self-organization of a wireless sensor network. *IEEE Personal Communications*, 7(5), 16–27.

Chapter 9

Energy-Aware Routing Protocol for Nanosensor Networks

Fadi Al-Turjman[1] and Kemal Ihsan Kilic[2]

[1]Department of Computer Engineering, Antalya Bilim University, Antalya, Turkey
[2]Department of Computer Engineering, Middle East Technical University
North Cyprus Campus, Kalkanli, Güzelyurt, Turkey

Contents

9.1   Introduction 151
9.2   Related Work 153
9.3   System Models 156
9.4   LaGOON Routing Protocol 159
9.5   Performance Evaluation 165
        9.5.1   Simulation Setup 166
        9.5.2   Performance Metrics and Parameters...166
        9.5.3   Simulation Results 167
9.6   Conclusion 173
References 173

9.1 Introduction

Internet of Things (IoT) is the next revolution after the Internet. With this technology, smart devices/things can communicate with each other. By using sensor devices connected through networks, our awareness about the environmental conditions

151

increases significantly. Particularly precise measurements from nanosensors are vital for the sustainable development of green systems. With the IoT (and especially with nanosensor networks) not only can energy statistics be collected but numerical data that provides information about the totality of the condition of the earth can be acquired. Applications such as air pollution control and measuring bio-degradation can utilize nanosensors for making finer measurements for a smarter and greener world. In the field of implanted wireless body area networks (WBANs), nanosensors play a vital role in monitoring the health of the human body. Furthermore, nanonetworks have opened new horizons to make an assessment of the conditions of the world at finer levels, with the help of communication at nanoscale.

However, at nanoscale, there are many design challenges. For the nanosensor nodes, revolutionary changes are required at the hardware and software level. The THz communication at the nanoscale imposes further restrictions over the range of the transmissions, and necessitates the multi-hop forwarding in nanosensors networks [1, 2]. This constraint makes routing vital for nanocommunication. On the other hand, the power unit and energy harvesting unit at nanoscale cannot sustain sensor nodes very long. For this reason, the energy-aware routing is a crucial design challenge that has to be addressed. In [3], THz communication and nanoscale energy harvesting are cited as the two important characteristics of the design of wireless nanosensor networks (WNSNs hereafter).

For this study, the focus was on the energy-aware routing protocols for the ad-hoc WNSNs. Considering the limitations of the WNSNs and the high energy cost of communication, a simple energy-efficient routing protocol, "LaGOON" (last good neighbor), is proposed. Keeping the lowest communication overhead and using the simplest data structure have been the two main principles that guided the design of the proposed routing protocol. The basic idea of the protocol was to make use of the information in the arriving packets as much as possible, with the assumption of symmetric communication. The adaptive "backward-learning" paradigm utilizes the path length information and the source of the packet for further future transmissions. Moreover, the method does not assume any specific topology. Since it involves an evolving distributed scheme, the proposed method can handle both mobile and static configurations in communication systems.

The proposed method is benchmarked against flooding and random routing. Flooding and random routing are two simple and widely used protocols. Flooding, being an instance of the multicast routing protocols, is computationally the simplest routing protocol in the sense of requiring no heuristics. But it is very costly when energy consumption is considered. The basic idea is to increase the chance of successful transmission and network coverage through the multiplication of packets, in which all the possible paths are traversed, including the shortest path (between the source and the destination). Periodic flooding is a technique that is used for exploratory purposes, like in [4]. Random routing is another computationally "light" routing method, in which a neighbor is chosen randomly to forward packets. For this reason, it can be regarded as the simplest unicast routing protocol.

This type of routing works well if the network is highly interconnected and does not lead to high energy consumption like flooding. Random forwards, in this case, can allow packets to reach the destination quickly if the shortest path is traversed by the packets. Also, random routing is proven to balance sensors' energy consumption in the wireless network (fair), as the next nodes are chosen randomly to transmit data [5,6]. In this regard, the LaGOON protocol is benchmarked against flooding, which is a computationally lightweight and robust protocol, and against random routing, which is a computationally lightweight and energy-wise "fair" protocol.

For the rest of this chapter, the breakdown of the sections is as follows: In Section 9.2, a general overview and related work to the nanosensor technology is given. Routing, the central topic of the chapter, is presented. Also, a short survey about the existing routing protocols is provided. After that, in Section 9.3, system model and network architecture that are assumed for WNSN are introduced. Peculiarities associated with WNSNs and their effects on the design are discussed. In Section 9.4, the proposed routing protocol, LaGOON, is explained and detailed. In Section 9.5, the simulation setup (Section 9.5.1), metrics and parameters that are used for the assessment (Section 9.5.2), and the results (Section 9.5.3) from the simulations are given. Finally, in Section 9.6, a qualitative assessment and comments are provided for the proposed method, and suggestions on improvements for the future work are discussed.

9.2 Related Work

Routing has a special stature in the WNSNs. Not only are WNSNs new technologies, but they are also fundamentally different technologies. Although there is no standard routing protocol for WNSNs yet, several promising studies are published. Unfortunately, current network protocols cannot be applied to nanocommunication. They are too complex for nanosensors, and the energy constraints of nanosensors cannot sustain such protocols. Due to the nanoscale dimensions of the sensor nodes and the size of their antennas, THz frequency is the only band where the communication can be done. As a result, new protocols, new modulation techniques, and new signal-encoding methods are necessary for THz frequency communications. The only work toward practical standardization is the framework IEEE 1906.1 [7], which contains recommendations. However, there are several studies focused on different aspects of the nanocommunication, like physical layer, THz antenna technology, routing, and energy harvesting. Several aspects can be considered in the design of a routing protocol such as the nanonetwork topology, multi-phase settings, mobility, and 2D-3D. But, according to current nanotechnology studies such as [1,3], the most important and limiting factor is the energy. In that sense, the special case of routing protocol design for the WNSNs can be seen as an energy optimization problem with various constraints.

Studies like those in the following paragraphs tried to develop routing frameworks focusing on different layers of the network. However, a comprehensive routing framework not only requires cross-layer design, but also collaborative involvement of various engineering departments. The ideal optimized routing requires synergy among various parts of the sensor node architecture such as the antenna, signal encoding method, duty-cycling mechanism, routing software, customized packet format, addressing scheme, and power unit controller. On the other hand, such efforts are the first steps towards standardization. In this sense the performance comparison among these methods not only can be difficult, but also cannot fully explain the isolated algorithm performance.

Authors in [8] focused on the physical layer part for their routing protocol. A physical network coding based on a pair-to-pair routing protocol was proposed. The packets are divided into two parts and transmitted in pairs along the pipelined multi-hop route, with the idea of gathering weak nodes into groups where integrated routing can be achieved for energy efficiency. The paper extends the idea of greedy perimeter stateless routing (GPSR) in [9]. While the GPSR is a point-to-point routing algorithm, authors extend it to be a pair-to-pair routing algorithm in their work, calling it a "buddy unicast" routing. Although the buddy unicast routing is not energy efficient, its performance in throughput is very high.

In [3], the authors, focusing on a routing framework for WNSNs, proposed a method for optimizing the harvested energy for the perpetual working of the network. The proposed framework tried to increase the network throughput together with the energy-harvesting optimizations. The central idea for the routing framework was hierarchical clustering. In doing so, the proposed routing method was to transfer the operation load from sensor nodes to cluster heads (nanocontrollers). However, the routing framework proposed is based on the special MAC protocol, which was the previous work of the authors. The paper presented simulation results of the proposed method by specifying the energy-saving probabilities of the sensor nodes depending on the distance between the sensor node and the cluster head (nanocontroller). The proportional relationship between the energy-saving probability and the distance between the nanosensor node and the cluster head was listed among the findings of the paper. However, the routing framework proposed was not compared with any other specific routing method.

In [10], a channel-aware routing protocol is proposed. The authors considered the special attributes of the THz band communication. The forwarding is optimized by considering two cost factors: namely, avoiding long-distance region in which the signal may suffer the path loss, and avoiding the short-distance region in which the number of hops may be increased unnecessarily. For the simulations a linear network topology is considered. The paper focused on metrics like end-to-end performance, end-to-end capacity, and end-to-end delay. For the parametric simulations, authors used different node densities and different water percentages representing various human body tissue characteristics, as the targeted application was the in-body health monitoring system. The proposed

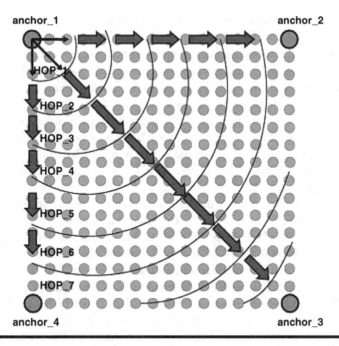

Figure 9.1 The anchor nodes-based coordinate system [11].

routing algorithm achieved high channel capacity and exhibited low hop count reducing transmission delays.

In [11], a coordinate-based addressing scheme, CORONA, is proposed for nanonodes placed uniformly in a rectangular 2D topology. The proposed routing protocol tries to minimize the hop count of the packet transmission by placing anchor nodes at the vertices of the grid and using the coordinate-based addressing scheme, as can be seen in Figure 9.1. Through simulations, the routing protocol is assessed by considering the packet retransmission rate, successful packet reception rate, and packet loss rate. According to the simulations done in the paper, CORONA showed low packet retransmissions and low energy consumption compared to the flood routing. However, the paper stated that global packet reception rate is very low for CORONA. Although the paper does not present an energy efficiency measure, the energy efficiency is associated with the metrics that are used in the simulations, such as send, receive, and interference rates. The CORONA protocol is benchmarked against DIF (the dynamic infrastructure) in [12] and probabilistic flooding [13]. The DIF protocol suggested in [12] can be regarded as an improvement over the plain flooding for preventing a "broadcast storm" problem.

In [14], authors proposed peer-to-peer type routing protocol. For the simulations, 2D uniform grid and 2D uniform random topologies are assumed, in which identical nanonodes are deployed. The assumption for the nanocommunication was 100GHz frequency with standard atmospheric conditions. This condition is assumed for providing the minimum effect for the path loss due to molecular absorption and high data rate. Packet collisions and redundant retransmissions, being the two metrics that are considered in the paper, are optimized in the proposed protocol. During the deployment phase, nodes are classified based on the packet reception statistics they have logged. The routing scheme exploits this classification.

The paper [15] proposes an energy-conserving protocol based on the hybrid clustering of the nanonodes and centralized scheduling. The proposed method, the nanocluster composition algorithm (NCCA), offers a model designed for channel behavior, by considering aggregated impact of molecular absorption, spreading loss, and shadowing. The energy model of NCCA considers harvesting in addition to the consumption.

In [16], the authors pointed out the lack of research related to the protocols in higher network layers for the WNSNs. They also added that most research is focused on the lower network layers, especially on the MAC layer and physical layer. After benchmarking classical WSN protocols like ad-hoc on-demand distance vector (AODV) [17–19], destination sequenced distance vector (DSDV) [20], and dynamic source routing (DSR) [21], on WNSN, authors found out that AODV performs better in WNSNs. Their performance metric criteria include packet delivery ratio, throughput (Kbps), average delay, packet drop rate, and energy consumption. As a parameter, the paper in [16] proposes a varying number of nanonodes (50, 100, 150, 200) in order to simulate sensor networks with varying densities, from sparse to dense. The paper based on this benchmark, proposed a modified version of AODV as a hierarchical AODV routing protocol. The protocol, which is customized for WNSNs, performed better than all the others. The paper [22] provides benchmarks for the current nanocommunication routing protocols. In the benchmarks, three routing protocols were considered: namely, controlled flooding, coordinate routing (CORONA [11]), and hierarchical AODV [16]. The performance metrics were focused on the number of successfully delivered packets, delay, and energy consumption. Authors used parameters such as the number of nanonodes (50–250 nodes, sparse-to-dense network spectrum) and their transmission range (1 mm, 10 mm, 15 mm, 20 mm). CORONA and controlled flooding are found to be worst in delay and energy consumption for increasing numbers of nodes and transmission ranges. While the hierarchical AODV protocol is found to have the best energy consumption, it presented higher complexity and lower throughput.

In Table 9.1, existing approaches related to the WNSNs are listed and compared.

9.3 System Models

In today's technology, nanosensors are utilized in various domains like biomedical, industrial, environmental, and military applications. With the networking technology, nanosensors, and in general nanomachines, become more potent, since they can cooperate and communicate to achieve more challenging tasks. Figure 9.2 shows the

Table 9.1 Comparison of Existing Routing Algorithms for WNSNs (as of 2018)

Paper	Contribution Summary
Zhou 2012 [8]	PHY layer and pair-to-pair routing. Not very energy efficient.
Pierobon 2014 [3]	Customized clustering for energy harvesting. Needs special MAC layer.
Yu 2015 [10]	Channel-aware routing protocol. 1D topology. Energy not considered.
Liaskos 2015 [11]	CORONA. Minimize hop count. 2D Grid topology. Energy considered. "Anchor" nodes.
Liaskos 2016 [14]	Peer-to-peer routing. 2D Grid topology. Node classification based on past statistics.
Tairin 2017 [16]	Hierarchical AODV. Energy considered.
Afsana 2018 [15]	Channel aware energy conserving protocol. Hybrid clustering of the nanonodes and centralized scheduling.
Proposed method	Lightweight energy aware protocol. Minimize hop count. Topology independent.

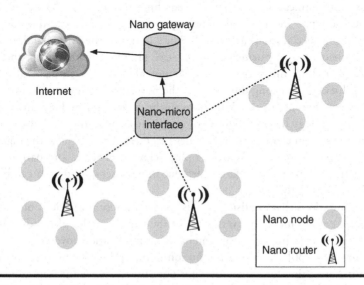

Figure 9.2 Network architecture and components for nanonetworks.

Table 9.2 Comparison of Communication Technologies Used in Nanonetworks

Comm. Type	Internet	Nano-Molecular	Nano-Wireless
Signal Type	EM	Chemical	EM
Signal Speed	High	Low	High
Power Consumption	High	Low	Low

general network architecture to be assumed in this chapter for the targeted nanonetworks. Important elements of the nanonetworks are nanonodes, nanorouters, and nano-micro interfaces. Through gateways, these types of networks can be connected to the traditional Internet.

The communication in nanonetworks can utilize one of the following technologies; nanomechanical, acoustic, electromagnetic and chemical or molecular communication [23]. A comparison for the existing communication technologies can be summarized as shown in Table 9.2, where EM stands for electro-magnetic. In the table, it should be noted that the comparisons are made based on the relative values among the communication types. However, for the energy consumption, if the comparisons are made by considering the ratios to the power unit capacities, nanoscale electromagnetic communication would require higher ratio than the Internet.

Mainly due to their tiny sizes, nanonetworks introduce difficulties in both hardware and software design. Detailed explanations of the hardware design of the electromagnetic wireless nanosensors can be found in [1]. Especially in the software part, the communication layer stack needs fine tuning as the hardware imposes many restrictions. The physical signaling is at the THz levels, due to antenna size, requiring special modulation techniques [24]. On the other hand, promising research is being done by using graphene-based plasmonic materials for antennas to overcome signaling difficulties [24,25]. Nanoscale batteries, in particular, cannot store much energy for long durations of the non-stopping operations in the nano nodes. Mostly because of the limitations in the current technology, any design on the nanotechnology has to follow a bottom-up approach. The structure of the system and the system element in nanoscales determine the design of the higher-level elements. This is shown in Figure 9.3 [26] below. Innovative and nature-inspired ideas related to topology and fault tolerance in nanonetwork design can be found in [27], [28], and [29].

The limitations that are listed in this section impose restrictions on the design of network communication layers. One of the best examples of the difficulties in software is the routing protocols for nanocommunication (overview on routing can be found in [30]). Considering the communication as being the most energy-consuming operation for nanonodes, it is not difficult to see the importance of the energy efficient methods in communications. The energy cost of a routing protocol

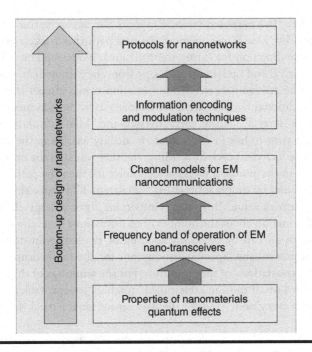

Figure 9.3   Bottom-up approach to the design of nanonetworks [26].

is high for a nanodevice, and nanobatteries are very limited. In fact, effectiveness of routing plays important role in energy-efficient use of nanonetworks. Differing from the traditional network routing protocols, nanorouting protocols have to be energy-aware optimization methods. Unlike the regular TCP/IP communication where devices can emit thousands of packets in seconds and the buffer memory is not a problem, in nanocommunication only a few packets in a minute can be sent and the memory is limited.

9.4 LaGOON Routing Protocol

As the nanodevices have simple hardware, rather restricted software is needed. The layered structure for the TCP protocol stack was proposed to overcome complexities of big software systems. However, for the nanodevices, software should be rather simple and efficient. The routing protocol should not introduce unnecessary communication overhead and has to be energy efficient. With these two principles in mind, this chapter proposes simple but energy-aware routing protocols for WNSNs.

The proposed method integrated deployment phase into the actual communication. There is no phase solely dedicated to the deployment. The topological information and the optimum routing evolve with time as the nodes communicate with each other. The proposed method exploits the duplex nature of the wireless transmissions. The idea is to remember the best path that packets arrives on, and use it in reverse direction for future transmissions. Nodes update information on every packet arrival and update the path cost from their immediate neighbor to the originator of the arrived packet. This information comes almost freely with every packet. No additional bytes or attribute is necessary to be allocated for this extra information, as "layer 3" (end-to-end addresses) and "MAC" (link-based) addresses along with the time to live (TTL) values are already written in the packets. Here, TTL value is deducted on each forwarding. In other words, if the max value is kind of hard-coded in the protocol (every node knows it), then the path-length or hop count can be found by simply subtracting the current TTL value that comes with a packet from the max value. This value is important for the energy-efficient routing, as it gives information about the number of transmissions necessary for the packet. Initial transmission can be based on flooding or random point-to-point routing with some improvements. Specific assumptions for the implementation are given below in the explanations of the algorithm. For the workings of the method, each node will have a small routing table called "reachability-cost-table." For routers it can be full (for every node) and for the nanosensor node partial (fixed number of entries).

The table will consist of triplets (src, cost, dst-neighbor):

- **src** is the originator (source) of the packet.
- **cost** is the total energy cost required to send the packet from src to dst.
- **dst-neighbor** (dn) is the last node on the path before the dst, the immediate neighbor of the destination node. This will be determined by the destination node.

Figure 9.4 illustrates the nodes that are mentioned in the discussion above. Assuming duplex transmission, then the "dst" node can mark in its table the cost to reach to "src" via "dst-neighbor" as "cost." This is the information extracted for the "Backward" path shown in Figure 9.4. Any value less than that updates the table of the node. In that way table entries will tell the nodes the best neighbor or the last good neighbor (LaGOON, "dn" in Figure 9.4) to pick for transmitting back to source node in the future. The assumption that each node knows its neighbor must be considered along with the algorithm, in order to fully understand operation of the routing protocol. For simulation purposes, when the packet is created randomly by a node, this node chooses a random destination and a random neighbor to initiate the transmission process. In our implementation, an unreliable (without ACK) transmission scheme is used. A reliable transmission scheme can be implemented by adding an ACK mechanism and the information that comes from the ACK packet can be helpful to update the "Forward" (Figure 9.4) path information as it is proposed in [31]. Although dynamic topology is assumed in the implementation, the nodes are confined to rectangular regions so that neighbors will not be outside

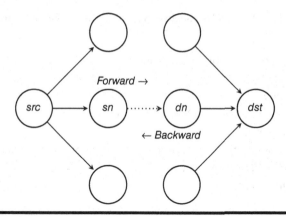

Figure 9.4 Illustration of the special nodes. *sn* **and** *dn* **are src-neighbor and dst-neighbor, respectively.**

of the range. The Gauss–Markov mobility model that is used in the sample health-care application of the Nano SIM package is assumed in our simulations.

The pseudocode of the proposed method is given in Algorithm 9.1. The algorithm is used by every node to decide on forwarding. Each node is either in "FULL" state or in "CHARGING" state, depending on the battery level.

- For step 1, every node initializes its routing table by marking all the distances to other nodes as NOT-REACHABLE. With the arrival of the first packet, nodes start to utilize the information that comes with the packet. As they receive packets that come from the other nodes, the TTL field gives an idea about the path length (hop-count) for the backward path. If a shorter path is found, the nanosensor node updates the last good neighbor and the path length to it, which is a simple cost metric.
- The nodes that are in "CHARGING" state basically "DROP" the packet since there is not enough power to "FORWARD," but use the information ("UPDATE" the routing table) that comes with the packet. It is assumed that nodes put themselves into "CHARGING" state when there is "low" (predetermined value) level of energy left. This process is summarized in steps 2–4 of Algorithm 9.1.
- The steps from 5–15 are related to the activities of the nodes which are in "FULL" state.
  - The nodes that have enough battery for forwarding firstly check the TTL value of the packet. Packets with expired TTL are "DROP"ed (step 6–7).
  - Packets that are destined to the node (node is "dst" node as it is shown in Figure 9.4) are "PROCESS"ed (step 8–9).
  - For the packets that are forwarded to the intermediate nodes on the "src-dst" path (please see Figure 9.4), the "FORWARD"ing is applied

Algorithm 9.1   LaGOON Routing Mechanism for Each Nanodevice

Input: Packet Arrival

Output: Drop, Forward, or Process Packet

Initialization:

1 Setup routing table:

   Mark all other nodes as NOT-REACHABLE

   On Packet Arrival:

2   UPDATE routing table according to the information in the packet

3   **if** (*STATE* == *CHARGING*) **then**

4   DROP Packet

5   **else**

6   **if** (*TTL* == 0) **then**

7   DROP Packet

8   **else if** (*LAYER3*$_{DST\,ADDR}$ == *ID*$_{NODE}$) **then**

9   PROCESS Packet

10   **else if** (*MAC*$_{DST\,ADDR}$ == *ID*$_{NODE}$) **then**

11   FORWARD Packet to *LAYER3*$_{DST\,ADDR}$

            via "LaGOON". Flood/Random forward if no "LaGOON" in the table

12   **else**

13   This is "RUMOR PACKET", do nothing

14   **end if**

15   **end if**

---

according to the most recent state of the routing table that the intermediate node has. As it is the case for every packet, addresses and TTL values in the packet are valuable information and before forwarding takes place, they are used to "UPDATE" the routing tables of the intermediate nodes, which are summarized in step 10–11 of Algorithm 9.1.

– Because of the wireless nature of the transmissions, nodes can also "listen" to other packets that are not destined to them. These packets are labeled as "RUMOR" packets in step 13 of Algorithm 9.1. These packets are also valuable to "UPDATE" the routing tables of the "hearing" nodes.

Many improvements can be suggested to LaGOON. As the "Backward" path discovery is explained above, it is also possible to do "Forward" path discovery (please see Figure 9.4) from the arriving packets. The immediate neighbor of the "src" ("sn" in Figure 9.4) can be included in the message of the packet (piggybacking) and the destination node ("dst" in Figure 9.4), depending on the energy level, can send back some kind of "ACK" packet to the originator ("src" in Figure 9.4), informing the cost of the transmission via "sn." This could help originator nodes to update their routing table for the cost of the "Forward" path. In this case, packet traffic can increase very much. To alleviate this problem, some kind of "probability of sending" value can be integrated to the protocol, as it is in the case of "persistence policies" of the CSMA. The basic idea is to use the communication packets as the source of information to determine the best neighbor for a specific destination and also inform the originator node about the path cost.

For visualizing the operation of the algorithm and the use of the associated data structures, Figures 9.5–9.11 are given below. Figures 9.5–9.8 show sample transmission from node "a". In this scenario, it is shown when the data transmission from node "a" happens towards node "g" via node "d" it takes more time than the transmission from node "b". In case the transmitted data reaches earlier to "g" via "d", the less costly path is not affected by the arrival of the packet via "b" since its cost is higher than the one registered in the routing table. The global routing table (for this example) is shown next to the network graph to show the individual updates to the routing tables that nodes keep track of for themselves. In the actual implementation, each node, as time passes and as they receive packets (from transmissions and from floods), updates its individual routing table. Figures 9.9–9.11

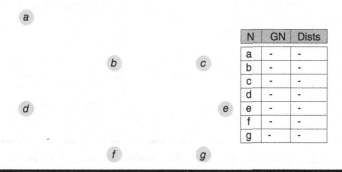

Figure 9.5   Initial collection of entries from routing tables of nodes.

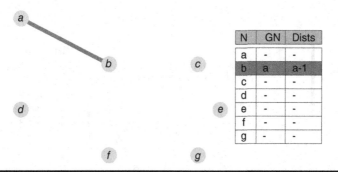

Figure 9.6   In this case, node "a" has an empty routing table and it starts the data transmission. When the data packet arrives to "b", it updates the fact of reaching "a" via "a" will cost 1, and it marks "a" as a neighbor.

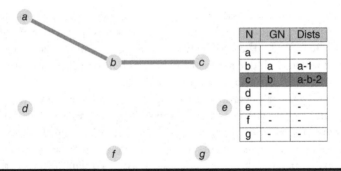

Figure 9.7 "c" updates the fact that for reaching "a" via "b" it will cost 2, and it marks "b" as a neighbor.

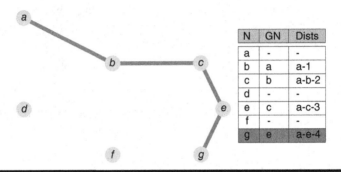

Figure 9.8 "g" updates the fact that for reaching "a" via "e" it will cost 4, and it marks "e" as a neighbor.

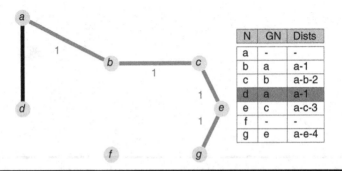

Figure 9.9 "d" updates the fact that for reaching "a" via "a" it will cost 1, and it marks "a" as s neighbor.

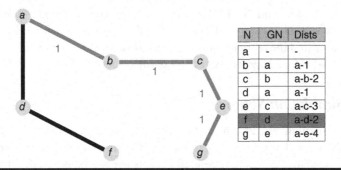

N	GN	Dists
a	-	-
b	a	a-1
c	b	a-b-2
d	a	a-1
e	c	a-c-3
f	d	a-d-2
g	e	a-e-4

Figure 9.10 "f" updates the fact that for reaching "a" via "d" it will cost 2, and it marks "d" as a neighbor.

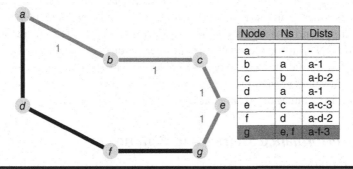

Node	Ns	Dists
a	-	-
b	a	a-1
c	b	a-b-2
d	a	a-1
e	c	a-c-3
f	d	a-d-2
g	e, f	a-f-3

Figure 9.11 "g" updates the fact that for reaching "a" via "f" it will cost 3, and it marks "f" as a neighbor.

show how the algorithm updates data structures when a better alternative route is found between source and destination. In routing tables, "N" is the node and "GN" represents the "last good neighbor." The "Dists" column represents the cost of the path in terms of the hop count. If a node has an "empty" routing table, which is the case at the startup, it can be programmed to do flooding or to send to a randomly picked neighbor (if neighborhood discovery is possible to do beforehand). Flooding can be customized to be "directional," in which nodes that "flood" do not repeat the "echoes" of their transmissions.

9.5 Performance Evaluation

Simulations are performed in this section by using "ns-3" [32], which is a discrete-event network simulator for Internet-based systems. Specifically, we used the Nano-SIM [33] library in ns-3, which has been mainly developed for nanonetworks [34, 35]. The Nano-SIM package is modified to include "battery level" and two states, "FULL" and "CHARGING," at the nanosensor node. The proposed method, LaGOON, is compared through simulations with benchmarks such as the plain flooding and the random routing (nodes are chosen randomly for forwarding) in the Nano-SIM package, to see if there is any improvement.

9.5.1 Simulation Setup

For the simulations, the healthcare [35] example program of the Nano SIM package is utilized by changing some parameters. The simulation time is fixed at seven seconds. The system generated about 10 packets per second per node for the transmissions. Some of the packets are dropped prematurely if the sending node is in charging stage or simply does not have enough energy. Four different values for the transmission range of nanonodes are used as 0.005m, 0.01m, 0.015m, and 0.02m. The transparent MAC [34,35] scheme is utilized for simulations. The Nano-SIM package assumes that the nanosensor nodes already know their neighbors. The same assumption has been used in our simulations and in the implementation of the LaGOON routing protocol. The number of devices is fixed to be equal to 1 nanogateway, and 10 nanorouters. However, for the nanosensor nodes, four different count values are assumed as follows: 50, 100, 200, and 300. The summary of the important simulation parameters is given in Table 9.3.

9.5.2 Performance Metrics and Parameters

The LaGOON routing is benchmarked against the baseline cases of plain flooding and random routing. Four metrics are selected to assess the overall energy efficiency and the transmission efficiency: namely, the number of "CHARGED," "DROP," "SEND," and "FORWARD" events. To get the number of these events, 50 simulations are carried out and the average number of these simulations are rounded.

For the overall energy efficiency, the number of times a node experiences a "CHARGED" event has been taken into consideration and counted. This simple

Table 9.3 Summary of the Simulation Parameters

Parameter	Value
Simulation time	7 sec
Packet rate	10 packet per sec
Transmission range	0.005m, 0.01m, 0.015m, 0.02m
Number of gateways	1
Number of routers	10
Number of nodes	50, 100, 200, 300

measure shows indirectly the level of energy the system has used in the duration of the simulation. The consumed energy level and the packet traffic are proportional to the number of the "CHARGED" event. Energy-efficient routing should avoid redundant energy consumption, unless this helps to optimize further transmissions. Unnecessary "FORWARD" events may cause packets to go around and consume the energy of the devices. Packets that are forwarded many times may be dropped because of two reasons: expired TTL and arrival to the node in "CHARGING" state.

The number of "DROP" packet events is a simple metric that can measure the performance of the overall transmission.

The number of "SEND" events depends on mainly on the battery capacity of the nanonodes. Most of the packets are dropped prematurely, because when they are created to be sent, the associated nanonode selected by the simulation system was in "CHARGING" stage. Since this was valid for all routing algorithms, no adjustments were necessary, as comparison was required. The aim in measuring "SEND" events was to see roughly the fraction of time that is spent for the "CHARGING" stage by the nodes utilizing specific routing algorithms. Successful algorithms should yield higher numbers of "SEND" events. If the algorithm is spending too much time forwarding due to bad routing decisions, then nodes will be in "CHARGING" stage most of the time, and the algorithm will yield very low numbers of "SEND" events.

The number of "FORWARD" events basically tells how parsimonious the specific routing algorithm is in carrying out the routing. Fewer of "FORWARD" events are desired, as this means less energy is spent in carrying packets to their destinations. As a result, better energy-aware routing should have less "FORWARD," less "CHARGED," more "SEND," and less "DROP" events. The simulation time is fixed at seven seconds for all routing algorithms as it is stated.

Four different transmission range values are used for the nodes, as shown in Tables 9.4–9.7. The idea here is to see the performance of the overall system under the increased transmission traffic. As the transmission range of the nodes increases, the number of receiving nodes also increases. Especially in the case of flooding, this causes many copies of the packets to be generated for forwarding.

9.5.3 Simulation Results

The results are summarized for "DROP" events in Table 9.4 and Figure 9.12, "CHARGED" events in Table 9.5 and Figure 9.13, "SEND" events in Table 9.6 and Figure 9.14, and "FORWARD" events in Table 9.7 and Figure 9.15. In the 3D bar charts (Figures), the number of nodes and the transmission ranges are listed on the right horizontal axis. Four different levels of transmission ranges (0.005m, 0.010m, 0.015m, 0.020m) are permuted with four different number of nodes (50, 100, 200, 300), producing a total of 16 levels. Each metric listed in this section represents an average of 50 simulation results. For three different methods (flood, random, lagoon), four different levels of transmission range and four different number of nodes are used. In total, 2400 simulations are carried out.

Table 9.4 Comparison of "DROP" Events

| TX Range | Method | Number of Nano Nodes | | | |
		50	100	200	300
0.005	lagoon	716	1668	3599	5498
	random	858	2013	4114	6156
	flood	2603	8512	28394	54689
0.01	lagoon	853	1608	3474	5325
	random	1002	2127	4316	6546
	flood	4220	14119	40593	79041
0.015	lagoon	883	1524	3113	4797
	random	1043	2317	4626	7013
	flood	5955	17708	51303	89375
0.02	lagoon	900	1206	2589	4105
	random	1167	2524	5138	7867
	flood	6462	20492	65714	154610

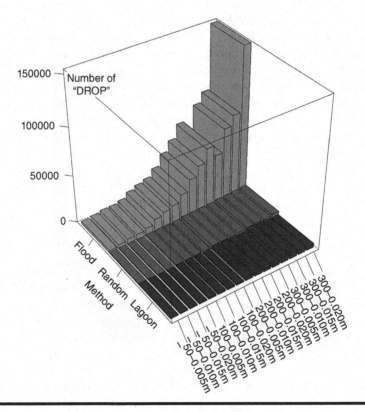

Figure 9.12   3D bar chart for the "DROP" events.

Table 9.5 Comparison of "CHARGED" Events

| TX Range | Method | Number of Nano Nodes | | | |
		50	100	200	300
0.005	lagoon	224	444	893	1336
	random	227	458	909	1340
	flood	266	569	1175	1770
0.01	lagoon	226	428	839	1252
	random	226	434	858	1283
	flood	286	593	1196	1796
0.015	lagoon	219	421	825	1226
	random	220	421	825	1226
	flood	295	601	1200	1799
0.02	lagoon	219	420	822	1222
	random	218	420	822	1222
	flood	302	603	1200	1803

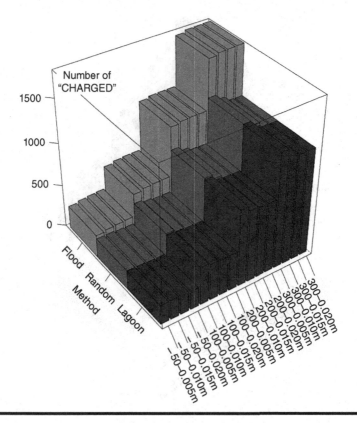

Figure 9.13   3D bar chart for the "CHARGED" events.

Table 9.6 Comparison of "SEND" Events

TX Range	Method	Number of Nano Nodes			
		50	100	200	300
0.005	lagoon	1152	2360	4604	6871
	random	1102	2260	4488	6607
	flood	757	1183	1572	1911
0.01	lagoon	1215	2542	4946	7328
	random	1231	2415	4728	7070
	flood	643	831	1124	1419
0.015	lagoon	1343	2790	5496	8131
	random	1319	2646	5150	7689
	flood	536	678	998	1194
0.02	lagoon	1463	2951	5900	8840
	random	1439	2888	5763	8626
	flood	445	599	828	1192

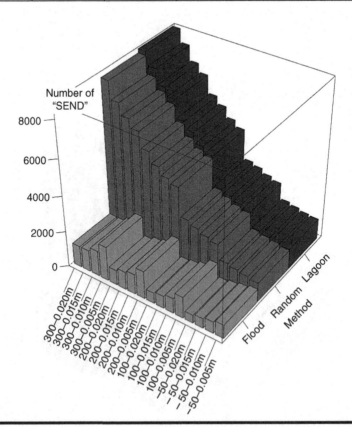

Figure 9.14   3D bar chart for the "SEND" events.

Table 9.7   Comparison of "FORWARD" events

TX Range	Method	Number of Nano Nodes			
		50	100	200	300
0.005	lagoon	749	1344	2771	4157
	random	793	1536	3002	4452
	flood	1472	3467	8026	12554
0.01	lagoon	713	1127	2231	3351
	random	688	1266	2478	3677
	flood	1702	4020	8648	13249
0.015	lagoon	550	820	1592	2442
	random	579	992	1981	2924
	flood	1892	4234	8811	13515
0.02	lagoon	402	511	938	1358
	random	425	701	1272	1859
	flood	2045	4331	8986	13536

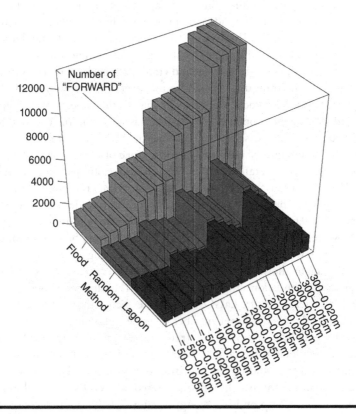

Figure 9.15   3D bar chart for the "FORWARD" events.

The number of "CHARGED" events are lower in LaGOON, which can be seen from Table 9.5. The implication of the results is that the LaGOON has better energy efficiency and better transmission efficiency compared to the plain flooding protocol. Considering Algorithm 9.1, it can be seen that the space complexity is at the order of $O(N)$ where N is the number of the nodes. Whereas the time complexity to select the best neighbor is just $O(1)$. Because a fast table lookup mechanism can be used while utilizing the nanonodes IDs. Considering computational cost, this is not much different than picking a random node to forward. It can be said that the difference between LaGOON and random routing is insignificant. But when other factors are considered, like the number of "DROP" events, the LaGOON is better than random routing. In fact, LaGOON has the lowest "DROP" rate, and it is the most reliable protocol compared to flooding and random routing.

This difference can be seen in Table 9.4 more clearly, where the number of the "DROP" is very low in LaGOON protocol compared to the plain flooding protocol and lower than the random routing. This is expected, as the LaGOON is keeping the minimum number of forwarding events. LaGOON is also superior to random routing. The difference between LaGOON and random routing is less than the difference between LaGOON and flooding. But from Table 9.4 it can be seen that for increasing range and number of nodes (denser traffic), the difference between LaGOON and random routing grows. This means denser traffic can help the random routing approach in finding alternative routes toward their destination. But it also increases the chance of choosing charging nodes and circulating packets unnecessarily longer in the network. One thing that should be noted here is the number of "DROP" events for the flooding protocol, since many copies of the same packet are generated and forwarded. But still, drop rate for the packets is so high in the case of flooding.

Table 9.6 shows the statistics about number of "SEND" events. This measure is related to the overall energy utilization performance of the protocol. It also shows the throughput of the overall system in terms of successful packet generation and delivery. If nodes are in charging state most of the time, the number of "SEND" events will be low. The only reason for this is bad routing decisions. As it can be seen from Table 9.6, LaGOON and random routing are almost similar in performance. However, without looking at the Table 9.7, it won't be easy to understand the difference between LaGOON and the random routing approach.

Table 9.7 again shows increasing difference between LaGOON and random routing as the transmission range gets wider and the number of nodes increases. Basically, for increased traffic LaGOON becomes better than random routing. In other words, in the case of lower number of nodes, choosing nodes for forwarding randomly may help, but when the traffic increases, this way may be costly, as explained in the previous paragraph. That is where LaGOON excels, by utilizing the minimalist way and adding few heuristics to routing decisions.

In general, in all the benchmarks, proposed LaGOON protocol performed very well, without compromising energy saving, reliability, and computational cost for routing decisions. Our results are justified via statistical tests which have been applied to verify the differences in mean values. One of the highlights of the LaGOON protocol is the fact that having such a good level of efficiency with a little computational cost can give space to further optimizations.

Flexibility is another desired feature in the proposed protocol. This protocol is not only topology independent, but also can be customized for handling different mobility models. Since the considered nanosensors are supposed to be distributed in dynamic human bodies which are in continuous movement status, mobility is a critical issue in this study. Node failures and node mobility have been resolved in this study by periodical control packets. These control packets are added easily for broadcasting the nanosensor position changes. This results in a faster route discovery. The "evolving learning" mechanism for route discovery is carried out in a distributive manner. The proposed LaGOON protocol can be customized to initiate routing with flooding or by choosing a random neighbor. Additionally, a hierarchical position information can also be added by imposing the required restrictions in the targeted application.

9.6 Conclusion

This chapter proposed LaGOON routing protocol as a simple and energy-aware routing protocol for WNSNs. Simulation results are presented to show the relative performance over the baseline cases (flooding and random routing). Improvements over the plain flooding and random routing suggest that the LaGOON protocol has potential uses in real life scenarios. Although the LaGOON protocol is presented without any enhancements, several improvements can be done to make it more efficient. One of the improvements, as mentioned at the end of Section 9.4, is the inclusion of the neighbor selected by the source node in the message, for optimization of the "FORWARD" path. In addition to this improvement, nanosensor nodes can decide to send a few preset probability parameters, similar to persistence schemes of the CSMA protocols. Another improvement can be done for the routing tables. While routers can store the whole routing table (for each node), nanonodes can use caching-type storage schemes to optimize the routing table size.

References

1. Akyildiz I.F., Jornet J.M. 2010. Electromagnetic wireless nanosensor networks, *Nano Comm. Networks*, 1(1), 3–19.
2. Canovas-Carrasco S., Garcia-Sanchez A.-J., Garcia-Haro J. 2018. A nanoscale communication network scheme and energy model for a human hand scenario. *Nano Communication Networks*, 15, 17–27.
3. Pierobon M., Jornet J.M., Akkari N., Almasri S., Akyildiz I.F. 2014. A routing framework for energy harvesting wireless nanosensor networks in the terahertz band. *Wireless Networks*, 20(5), 1169–1183.

4. Intanagonwiwat C., Govindan R., Estrin D., Heidemann J., Silva F., 2003. Directed diffusion for wireless sensor networking. *IEEE/ACM Trans. Netw.*, 11(1), 2–16.

5. Nguyen M.T., Teague K.A. 2017. Compressive sensing based random walk routing in wireless sensor networks. *Ad Hoc Networks*, 54(supplement C), 99–110.

6. Angelopoulos C.M., Nikoletseas S., Patroump D., Rapropoulos C., 2011. A new random walk for efficient data collection in sensor networks. In *Proceedings of the 9th ACM International Symposium on Mobility Management and Wireless Access,* 53–60.

7. IEEE. IEEE 1906.1-2015 - IEEE Recommended Practice for Nanoscale and Molecular Communication Framework. https://standards.ieee.org/standard/1906_1-2015.html, accessed 4-October-2018.

8. Zhou R., Li Z., Wu C., Williamson C. 2012. Buddy routing: A routing paradigm for nanonets based on physical layer network coding. In *2012 21st Int. Conf. on Computer Comm. and Networks (ICCCN)*, 1–7.

9. Karp B., Kung H.T. 2000. GPSR: Greedy perimeter stateless routing for wireless networks. In *Proceedings of the 6th Annual International Conference on Mobile Computing and Networking*, 243–254.

10. Hang Y., Ng B., Seah W.K.G., 2015. Forwarding schemes for em-based wireless nanosensor networks in the terahertz band. In *Proc. of the 2nd Annual Int. Conf. on Nanoscale Computing and Comm.*, 1–6.

11. Tsioliaridou A., Liaskos C., Ioannidis S., Pitsillides A. 2015. Corona: A coordinate and routing system for nanonetworks. In *Proc. of the 2nd Annual Int. Conf. on Nanoscale Computing and Comm.*, NANOCOM '15, 1–6.

12. Liaskos C., Tsioliaridou A. 2015. A promise of realizable, ultra-scalable communications at nano-scale: A multi-modal nano-machine architecture. *IEEE Transactions on Computers*, 64(5), 1282–1295.

13. Zhu T., Zhong Z., He T., Zhang Z. 2013. Achieving efficient flooding by utilizing link correlation in wireless sensor networks. *IEEE/ACM Transactions on Networking*, 21(1), 121–134.

14. Liaskos C., Tsioliaridou A., Ioannidis S., Kantartzis N., Pitsillides A. 2016. A deployable routing system for nanonetworks. In *IEEE Int. Conf. on Comm. (ICC) 2016*, 1–6.

15. Afsana F., Asif-Ur-Rahman Md., Ahmed Md.R., Mahmud M., Kaiser M.S. 2018. An energy conserving routing scheme for wireless body sensor nanonetwork communication. *IEEE Access*, 6, 9186–9200.

16. Tairin S., Nurain N., Al Islam A.B.M.A. 2017. Network-level performance enhancement in wireless nanosensor networks through multi-layer modifications. In *2017 International Conference on Networking, Systems and Security (NSysS)*. 75–83.

17. Hasan M.Z., Al-Rizzo H., Al-Turjman F. 2017. A survey on multipath routing protocols for QoS assurances in real-time multimedia wireless sensor networks, *IEEE Communications Surveys and Tutorials*, 19(3), 1424–1456.

18. Jhaveri R. 2015. Mobile ad-hoc networking with ADV: A review. *International Journal of Next-Generation Computing*, 6(3), 165–191.

19. Höfner P., Glabbeek R.V., Tan W.M., Portmann M., McIver A., Fehnker A. 2012. A rigorous analysis of AODV and its variants. In *Proceedings of the 15th ACM International Conference on Modeling, Analysis and Simulation of Wireless and Mobile Systems*, 203–212.

20. Perkins C.E., Bhagwat P. 1994. Highly dynamic destination-sequenced distance- vector routing (dsdv) for mobile computers. *SIGCOMM Comput. Commun. Rev.*, 24(4), 234–244.

21. Al-Turjman F. 2018. Wireless sensor networks: Deployment strategies for outdoor monitoring. Taylor and Francis, CRC: New York. ISBN 9780815375814.

22. Abuali N. *et al.* 2018. Performance evaluation of routing protocols in electromagnetic nanonetworks. *IEEE Access*, 6, 35908–35914.

23. Akyildiz I.F., Brunetti F., Blázquez C. 2008. Nanonetworks: A new communication paradigm. *Computer Networks: The Int. Journal of Computer and Telecommunications Networking*, 52(12), 2260–2279.

24. Jornet J.M. 2013. *Fundamentals of electromagnetic nanonetworks in the terahertz band.* Ph.D. thesis, School of Electrical and Computer Engineering, Georgia Institute of Technology, Atlanta, GA.

25. Jornet J.M., Akyildiz I.F. 2013. Graphene-based plasmonic nano-antenna for tera-hertz band communication in nanonetworks. *IEEE Journal on Selected Areas in Comm.*, 31(12), pp. 685–694.

26. Akyildiz I.F., Jornet J.M. 2010. The internet of nano-things. *IEEE Wireless Comm.*, 17(6), 58–63.

27. Akbas M.I., Solmaz G., Turgut D. 2016. Molecular geometry inspired positioning for aerial networks. *Computer Networks*, 98, 72–88.

28. Senel F., Younis M.F., Akkaya K. 2011. Bio-inspired relay node placement heuristics for repairing damaged wireless sensor networks. *IEEE Trans. on Vehicular Technology*, 60(4), 1835–1848.

29. Imran M., Younis M., Said A.M., Hasbullah H. 2012. Localized motion-based connectivity restoration algorithms for wireless sensor and actor networks. *Journal of Network and Computer App.*, 35(2), 844–856.

30. Neupane S.R. June 2014. Routing in resource constrained sensor nanonetworks. Master's thesis, Tampere University of Technology, Tampere, Finland.

31. Oteafy S.M.A., Al-Turjman F., Hassanein H.S. 2012. Pruned adaptive routing in the heterogeneous Internet of Things in *2012 IEEE Global Comm. Conf. (GLOBECOM)*, 214–219.

32. NS-3, "Network Simulator 3." https://www.nsnam.org, accessed 4-October-2018.

33. Nano-Sim, "NS-3 nanonetwork package." https://telematics.poliba.it/index.php?option=com_content&view=article&id=30&Itemid=204&lang=en, accessed 4-October-2018.

34. Piro G., Grieco L.A., Boggia G., Camarda P. 2013. Nano-sim: simulating electromagnetic-based nanonetworks in the network simulator 3. In *Proceedings of the 6th International ICST Conference on Simulation Tools and Techniques*, pp. 203–210.

35. Piro G., Grieco L.A., Boggia G., Camarda P. 2013. Simulating wireless nano sensor networks in the ns-3 platform. In *2013 27th International Conference on Advanced Information Networking and Applications Workshops*, 67–74.

Chapter 10

LCPC Code for Wireless Body Area Networks

Salah A. Alabady[1] and Fadi Al-Turjman[2]

[1]Computer Engineering Department, University of Mosul, Mosul, Iraq
[2]Department of Computer Engineering, Antalya Bilim University, Antalya, Turkey

Contents

10.1 Introduction 177
10.2 The Proposed LCPC Codes 180
    10.2.1 LCPC Code Encoding 180
    10.2.2 LCPC Decoding 184
        10.2.2.1 LCPC Error Detection 184
        10.2.2.2 Error Correction 188
10.3 Complexity Analysis 191
10.4 Simulation Results 193
10.5 Conclusions 199
References 200

10.1 Introduction

A wireless body area network (WBAN) consists of several tiny sensors that are located inside and outside the human body for continuous monitoring of vital parameters of patients suffering from chronic diseases. The wearable sensor unit consists of a transmitter, a receiver, and a central process unit called a gateway. The gateway is used to connect wearable sensors on the human body to the Internet.

In WBAN the network must deliver reports and patient health alerts in a perfect manner, in which delay or loss is not tolerated. To increase the lifetime of such networks, the energy spent by the sensors has to be minimized [1].

The implementation of WBAN in smart environments has recently been significantly affected by the emerging phenomena called Internet of Things (IoTs). IoTs and wireless sensor networks (WSNs) have attracted a lot of attention from researchers and have been applied in most aspects of our lives in various fields of technology [2]. The main idea of this concept is the ubiquitous presence around us of a variety of things or objects that are connected to the Internet, such as mobile phones, laptops, and daily-use objects like refrigerators, televisions, and smart sensors.

Mainly due to the rapid proliferation of wearable devices, smart sensors, and smartphones, the IoT-enabled technology is evolving from a conventional hub-based system to more personalized systems. Efficient IoT-enabled systems can be realized by providing fault-tolerant access to rich information with unobtrusive monitoring. Wireless communication links, which are rapidly prone to failures, are at the heart of this concept, and their development is a key issue if such a concept is to achieve its potential [3, 4]. One major concern in IoT and WSNs is energy conservation and consumption. The fundamental challenge for the realization of the IoT-enabling technologies is energy-constrained communication [5]. Therefore, avoiding or reducing the number of retransmissions is an important issue. Depending on the channel condition, there are different schemes to reduce energy consumption. One scheme is to find the optimal frame size. If the channel is good, bigger frames make more sense, as they will all go through with less overhead. While the channel conditions are not as good, smaller frames are better, as the probability of having a frame in error is lower with smaller frames. Another strategy to reduce the energy consumption is by using forward error correction mechanisms. Hence, we focus in this article on error correction aspects of wireless connections in the IoT era [2]. We propose low-complexity parity check (LCPC) codes that detect and correct single- and double-bit errors.

During the past decade, many error detection and correction code schemes (turbo codes, RS, BCH, and LDPC codes) have been investigated to increase the reliability of the wireless network systems in order to fulfill the quality of the data in a high-data-rate wireless network. Each of the designed codes has its own advantage to be used as the channel coding scheme in a communication system. Recently, there has been widespread research on LDPC codes, since it offers near-Shannon-limit performance [5].

Among the earliest error detection and correction codes available are the Hamming codes, which are able to detect up to double-bit errors but just able to correct a single-bit error. This constraint to correct double-bit errors can be attributed to a limited number of syndromes available. The codes consist of three rows and seven columns, which account for eight values of syndrome [5, 6]. In case of a single-bit error, there are seven possibilities of error patterns when the code word length equals seven bits, in which case each error pattern is assigned to one syndrome vector. On the other hand, double-bit errors throw 21 possibilities of error pattern for the same code word length equal to seven bits. The Hamming code does not have the requisite number of syndrome (i.e., 21), which results in each three error patterns being assigned by one syndrome vector. This makes the correction operation very difficult, and it fails to decide the correct error pattern from the three possible error patterns. In addition, Hamming codes cannot detect more than two-bit errors (e.g. burst error) [5, 7, 8].

RS code is another famous error correction code and is the subset of the BCH code [9]. For, a particular RS code specified as RS (n, k) with s-bit symbols, the number and type of errors that can be corrected in RS code depends on the characteristics of that code [10]. An RS decoder can correct up to t symbols that contain errors in a code word, where $2t = n - k$. To increase the capability of error correction, the number of the parity code must also increase. This means that the value of t in RS code must be very large.

Likewise, the error correction capability of the LDPC code also depends on the code word length and the characteristic of the parity check matrix [5, 6]. The error correction capability of the LDPC codes depends on the code word length and the characteristics of the parity check matrix [7, 8]. The decoder gives a better performance with a larger code word and with good parity-check matrices. In practice, to achieve a better BER performance with LDPC codes close to the channel capacity, the length of the LDPC code word used should be in the order of thousands of bits [9, 10]. The matrix multiplication for that big code word size demands huge memory, computational requirements, and more complex decoding [9, 11–13]. The LDPC codes fail to correct errors if the number of errors occurred is greater than the error correction capability of the decoder. Furthermore, LDPC codes require iteration in the detection and correction error processes around 10 to 50 times of iteration [14, 15]. Therefore, the need for efficient channel codes with less encoding/decoding complexity, and lower memory size requirement is quite obvious. For short code word-length coding, there have been several attempts in the literature. For example, authors in [16–21] take into consideration the short code word length for LDPC code. On the other hand, authors in [22–25] are interested in the short code word length for turbo codes.

This chapter considers various code rates for the LCPC code (i.e., 0.428, 0.375 and 0.444), LCPC (7, 3), LCPC (8, 3), and LCPC (9, 4). With the intention of producing a simple error correction code, in this work, we extend the capability of the Hamming code. The proposed LCPC codes offer lower encoding and decoding computational loads as compared to the Turbo code [25], RS, BCH, and LDPC codes, since the former does not involve an iteration process and requires very low memory and low complexity. Each LCPC code is represented as a short length code word, which makes the proposed codes particularly attractive for low-latency and real time applications of IoT and futuristic wireless networks applications.

The simulation results show that the proposed LCPC codes outperform other code types such as Hamming, RS, and LDPC codes. The LCPC (9, 4) code produces 3 dB coding gain while compared to the LDPC (8, 4) code at BER = 10^{-5}, and 1 dB at BER = 10^{-5}. On the other hand, LCPC (9, 4) code produces 2.1 dB coding gain as compared with the RS (7, 4) and 1.7 dB with the Hamming (7, 4) code at BER equal to 10^{-5}. The mean difference between LCPC code and Hamming code is that the latter can correct single-bit error and detect double-bit error, whereas, the LCPC codes can detect and correct single- and many cases of double-bit errors (i.e., 21 in case LCPC (9, 4)).

This chapter is organized as follows. Section 10.2 provides the description of the proposed approach to LCPC codes. Section 10.3 presents the complexity analysis of the LCPC code. Section 10.4 presents the performance of LCPC codes and analysis of simulation results. Section 10.5 contains the conclusions.

10.2 The Proposed LCPC Codes

In this section, a general proposed method for the LCPC codes is provided. The LCPC code is defined as block code (n, k), where n is the code word length and k the information length, respectively. In this chapter, three different kinds of LCPC codes are presented. In order to explain how the LCPC codes work, we present in detail the encoding and decoding of the LCPC (9, 4) code in the next subsections. The encoding/decoding of the LCPC (8, 3) and LCPC (7, 3) codes is identical with the LCPC (9, 4) code. Therefore, we only introduce **G** and **H** matrices of the LCPC (8, 3) and LCPC (7, 3) codes in addition to the error pattern and syndrome vector tables instead of showing the details.

10.2.1 LCPC Code Encoding

Figure 10.1 demonstrates a general block diagram of encoding and decoding processes in the LCPC code. The first step in the encoding process of the LCPC code is to segment the source data sequence into symbols of equal length (i.e., k bits). Subsequently, we take each symbol (k bits) and map it into code word c of n bits, where $n > k$. The $n - k$ additional parity-check bits are the redundancies added, which are used for error detection and correction. The LCPC code is a block code which takes the data stream from the source encoder, divides it into four-bit symbols (i.e., k), and then encodes each four-bit symbol (depending on the number of rows in **G** matrix) into a nine-bit code word (i.e., n) (depending on the number of columns in **G** matrix), before the transmission. The symbol of source data is denoted as $SD_i = (v_1, v_2, \ldots v_k)$, where $1 \leq i \leq j$, and j is the number of symbols of the source data, v is a binary bit, and $k = 4$ is the length of the symbol. Each k-bits symbol is then encoded into an n-bits code word before the

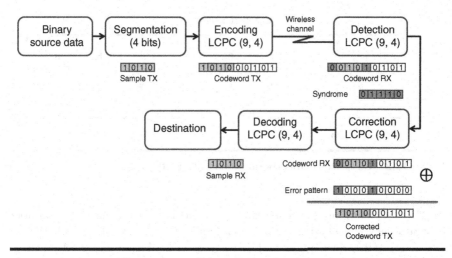

Figure 10.1   Components of encoding and decoding processes in LCPC codes.

transmission. The encoding process is implemented using the generator matrix **G**, which could be expressed as Eq. (10.1). The four left columns represent the identity (I) matrix and the five right column represent the parity (P) matrix. The parity (P) matrix is given in Eq. (10.4).

$$
\mathbf{G} =
\begin{bmatrix}
1 & 0 & 0 & 0 & 1 & 1 & 1 & 1 & 0 \\
0 & 1 & 0 & 0 & 1 & 1 & 1 & 0 & 1 \\
0 & 0 & 1 & 0 & 1 & 1 & 0 & 1 & 1 \\
0 & 0 & 0 & 1 & 1 & 0 & 1 & 1 & 1
\end{bmatrix}
\tag{10.1}
$$

In the encoding unit the redundant bits r is then added to each symbol to make the length of the code word equal to n, where $n = k + r$, and $r = 5$. The code word of the symbol corresponds to $CD_{Ti} = (\beta_1, \beta_2 \dots \beta_n)$, where $n = 9$, and β_i is a binary bit.

A code word CD_{Ti} is given by Eq. (10.2), used for encoding the symbols data that are defined as a multiplication between SD_i and **G**:

$$
CD_{Ti} = SD_i \times G
\tag{10.2}
$$

where, SD_i is an information symbol, CD_{Ti} is the transmitted code word, and **G** is the proposed generator matrix.

Figure 10.2 shows the code word construction of the LCPC (9, 4) code. The five bits ($\gamma_1, \gamma_2, \dots \gamma_5$) from the right side are the parity bits, while the four bits ($v_1, v_2, \dots v_4$) from the left side are the symbol bits. Eqs. (10.3) and (10.4) show the parity bits and symbol information bits respectively for LCPC (9, 4) code. In LCPC

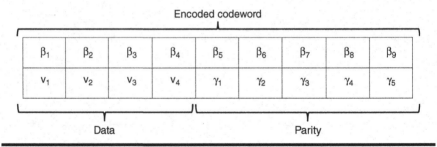

Figure 10.2   Code word construction for LCPC (9, 4) code.

(9, 4) code, the number of redundant parity bits is 5, so the maximum number of syndrome vector obtained is 32 (i.e., 2^5) and this means the LCPC (9, 4) code can detect and correct single- and many cases of double-bit errors. Meanwhile, in the Hamming (7, 4) code, the number of redundant parity bits is 3, so the maximum number of syndrome vector obtained is 8 (i.e., 2^3). In addition to the different **G** and **H** matrices in the LCPC (9, 4) code, the number of syndrome is also different.

The inability of Hamming code in correcting the double-bit error can be returned to the limited number of syndromes. In case of single-bit error, there are seven possibilities of error patterns when the code word length is seven bits. In this case, each error pattern is assigned one syndrome vector. In case of double-bit errors, there are 21 possibilities of error patterns, and the Hamming code does not have this number of syndromes (i.e., 21). Hamming code (7, 4) only has eight syndrome vectors (i.e., 2^m) and $m = 3$. Therefore, in case of double-bit error, each two or three error patterns are assigned one syndrome vector. This makes the correct operation difficult or impossible, because it cannot precisely detect the correct error pattern from the two or three error patterns. In addition, Hamming code cannot detect more than two-bit errors (e.g. burst error).

$$\beta_1 = v_1$$
$$\beta_2 = v_2$$
$$\beta_3 = v_3 \tag{10.3}$$
$$\beta_4 = v_4$$

$$\beta_5 = \gamma_1 = v_1 \oplus v_2 \oplus v_3 \oplus v_4$$
$$\beta_6 = \gamma_2 = v_1 \oplus v_2 \oplus v_3$$
$$\beta_7 = \gamma_3 = v_1 \oplus v_2 \oplus v_4 \tag{10.4}$$
$$\beta_8 = \gamma_4 = v_1 \oplus v_3 \oplus v_4$$
$$\beta_9 = \gamma_5 = v_2 \oplus v_3 \oplus v_4$$

The proposed parity check matrix **H** of the LCPC (9, 4) code is given by Eq. (10.5), which can be used in error detection.

$$
\mathbf{H} = \begin{bmatrix}
1 & 1 & 1 & 1 & 1 & 0 & 0 & 0 & 0 \\
1 & 1 & 1 & 0 & 0 & 1 & 0 & 0 & 0 \\
1 & 1 & 0 & 1 & 0 & 0 & 1 & 0 & 0 \\
1 & 0 & 1 & 1 & 0 & 0 & 0 & 1 & 0 \\
0 & 1 & 1 & 1 & 0 & 0 & 0 & 0 & 1
\end{bmatrix}
\tag{10.5}
$$

In a similar fashion, the proposed generator matrix **G** and parity check matrix **H** of LCPC (8, 3) and LCPC (7, 3) codes are represented by Eqs. (10.6–10.9). The encoding and decoding processes are the same for all LCPC codes.

G matrix for LCPC (8, 3)

$$
\mathbf{G} = \begin{bmatrix}
1 & 0 & 0 & 1 & 0 & 1 & 0 & 1 \\
0 & 1 & 0 & 1 & 1 & 0 & 1 & 1 \\
0 & 0 & 1 & 0 & 1 & 1 & 0 & 1
\end{bmatrix}
\tag{10.6}
$$

H matrix for LCPC (8, 3)

$$
\mathbf{H} = \begin{bmatrix}
1 & 1 & 0 & 1 & 0 & 0 & 0 & 0 \\
0 & 1 & 1 & 0 & 1 & 0 & 0 & 0 \\
1 & 0 & 1 & 0 & 0 & 1 & 0 & 0 \\
0 & 1 & 0 & 0 & 0 & 0 & 1 & 0 \\
1 & 1 & 1 & 0 & 0 & 0 & 0 & 1
\end{bmatrix}
\tag{10.7}
$$

G matrix for LCPC (7, 3)

$$
\mathbf{G} = \begin{bmatrix}
1 & 0 & 0 & 1 & 1 & 0 & 0 \\
0 & 1 & 0 & 0 & 1 & 1 & 0 \\
0 & 0 & 1 & 0 & 0 & 1 & 1
\end{bmatrix}
\tag{10.8}
$$

H matrix for LCPC (7, 3)

$$
\mathbf{H} = \begin{bmatrix}
1 & 0 & 0 & 1 & 0 & 0 & 0 \\
1 & 1 & 0 & 0 & 1 & 0 & 0 \\
0 & 1 & 1 & 0 & 0 & 1 & 0 \\
0 & 0 & 1 & 0 & 0 & 0 & 1
\end{bmatrix}
\tag{10.9}
$$

10.2.2 LCPC Decoding

The decoding algorithm consists of three parts. First part is the error detection that is implemented by computing the syndrome vector. Second part is the error pattern detection. Third part is the error correction.

10.2.2.1 LCPC Error Detection

The detection process detects errors in the received code word (CD_{Ri}), that is defined as the transmitted code word (CD_{Ti}) with errors (EP), as shown in Eq. (10.10). The parity check matrix **H** of LCPC code is used for this purpose. After the code word is received, the syndrome vectors (SY) are obtained from the (CD_{Ri}).

$$CD_{Ri} = CD_{Ti} + EP \tag{10.10}$$

$$SY = H \times CD_{Ri}^T \tag{10.11}$$

where SY = (γ_1, γ_2, ... γ_r) is the syndrome's binary vector. Therefore, Eq. (10.10), can be written as:

$$SY = H \times (CD_{Ti} + EP)^T \tag{10.12}$$

$$SY = H \times CD_{Ti}^T + H \times EP^T \tag{10.13}$$

since any row in the **H** matrix is orthogonal to the rows of the **G** matrix, and the inner product of a row in **G** with a row in **H** will be zero, the result of multiplication **H** by the CD_{Ti}^T is zero if there are no bit errors in the code word as shown in Eq. (10.14), where the CD_{Ti}^T is the transpose of the transmitted code word.

$$H \times (CD_{Ti})^T = 0 \tag{10.14}$$

From Eqs. (10.13) and (10.14), the SY can be expressed as:

$$SY = H \times (EP)^T \tag{10.15}$$

Here, EP^T is the transpose of the error pattern. From Eq. (10.15), it can be noticed that the syndrome vector SY depends only on the error pattern EP (i.e., number of bit errors and the position of the bit error). The error detection is implemented by calculating the SY value. For any received code word, if SY is the null vector (i.e., SY = 0), it indicates that the received code word (CD_{Ri}) is error-free. On the other hand, if SY is non-zero, there is change in bits, which means there are some bit errors.

Tables 10.1 and 10.2 show the EP and SY for LCPC (9, 4) code for single-bit error and the cases of double-bit errors in the received code word, respectively.

Tables 10.3–10.6 show the EP and SY for LCPC (8, 3) and LCPC (7, 3) when there is single bit-error and the cases of double-bit errors in the received code word, respectively.

Table 10.1 Error Pattern and Syndrome Vector for One-Bit Error of LCPC (9, 4) Code

Error Pattern EP	Syndrome Vector SY
000000000	0 0000
000000001	0 0001
000000010	0 0010
000000100	0 0100
000001000	0 1000
000010000	1 0000
000100000	1 0111
001000000	1 1011

Table 10.2 Error Pattern and Syndrome Vector for Two-Bit Errors of LCPC (9, 4) Code

Error Pattern EP	Syndrome Vector SY
000000011	0 0011
000000101	0 0101
000001001	0 1001
000010001	1 0001
000100001	1 0110
001000001	1 1010
010000001	1 1100
100000001	1 1111
000000110	0 0110
000001010	0 1010
000010010	1 0010
000100010	1 0101
001000010	1 1001
000001100	0 1100
000010100	1 0100
000100100	1 0011
000011000	1 1000
000110000	0 0111
001010000	0 1011
010010000	0 1101
100010000	0 1110

Table 10.3   Error Pattern and Syndrome Vector for One-Bit error of LCPC (8, 3) Code

Error Pattern EP	Syndrome Vector SY
00000000	0 0000
00000001	0 0001
00000010	0 0010
00000100	0 0100
00001000	0 1000
00010000	1 0000
00100000	0 1101
01000000	1 1011
10000000	1 0101

Table 10.4   Error Pattern and Syndrome Vector for Two-Bit errors of LCPC (8, 3) Code

Error Pattern EP	Syndrome Vector SY
00000011	0 0011
00000101	0 0101
00001001	0 1001
00010001	1 0001
00100001	0 1100
01000001	1 1010
10000001	1 0100
00000110	0 0110
00001010	0 1010
00010010	1 0010
00100010	0 1111
01000010	1 1001
10000010	1 0111
01000100	1 1111
00011000	1 1000
01001000	1 0011
10001000	1 1101
01010000	0 1011
01100000	1 0110
11000000	0 1110

Table 10.5   Error Pattern and Syndrome Vector for One-Bit Error of LCPC (7, 3) Code

Error Pattern EP	Syndrome Vector SY
0000001	0001
0000010	0010
0000100	0100
0001000	1000
0010000	0011
0100000	0110
1000000	1100

Table 10.6   Error Pattern and Syndrome Vector for Two-Bit Errors of LCPC (7, 3) Code

Error Pattern EP	Syndrome Vector SY
0000101	0101
0001001	1001
0100001	0111
1000001	1101
0000110	0110
0001010	1010
0010010	0001
0100010	0100
1000010	1110
0001100	1100
1000100	1000
0011000	1011
0101000	1110
1001000	0100
1010000	1111
1100000	1010

The number of error patterns can be computed using Eq. (10.16), where, n is the code word length (in the proposed code $n = 9$), and $e \in (1, 9)$ is the number of bit errors that may occur in the code words. In the proposed code, we assume the lookup tables that include the EP of each SY for single- and double-bit errors are stored in the memory.

$$NoEP = \frac{n!}{e!(n-e)!} \tag{10.16}$$

Once the code word the CD_{Ri} is received to the receiver side, the first step is to carry out error detection. In the error detection unit, the SY value is calculated using Eq. (10.11). The value of SY depends on the type of error (i.e., number of bit errors) and on the position of the bit error in the received code word CD_{Ri}.

Then, depending on the SY value, the EP can be determined from the lookup tables that are stored in memory. (e.g. Table 10.1 in case single bit error). Figure 10.3 illustrates the pseudocode for error detection and correction functions of the proposed LCPC codes.

10.2.2.2 Error Correction

In detection unit at the receiver side, for any CD_{Ri}, if the SY = 0, this means, there are no errors in the received code word, while SY/= 0 means there are some bit errors in the received code word (the error here may be in single-, double-, or more than double-bit errors; one can know that from the SY value). In the error correction process, the EP is chosen (depending on the SY value) and is fetched from the lookup tables that are stored in memory. The correction process is achieved as shown in Eq. (10.17).

$$\overline{CD_{Ti}} = CD_{Ri} \oplus EP \tag{10.17}$$

where $\overline{CD_{Ti}}$ is defined as the corrected transmitted code word after being received.

Figure 10.4 shows the flowchart for error detection and correction of single- and double-bit errors of the LCPC code at the receiver side. After the code word is received in the receiver side, the first step is to carry out error detection. In the error detection unit, the SY value is calculated using Eq. (10.11). The value of SY depends on the type of error (i.e., single- or double-bit errors) and position of the bit error in CD_{Ri}. Then, depending on the SY value, the EP from the lookup tables stored in memory can be determined.

As shown in Figure 10.4, if SY indicates that if there are single- or double-bit errors in CD_{Ri}, LCPC code can correct the received code word (CD_{Ri}). This correction is achieved by adding the specific EP to error received code word CD_{Ri}. Next, the decoding of the corrected received code word is carried out. The decoder is implemented by masking the last four bits on the left side of the code word.

Algorithm 1: LCPC Code Decoding

```
 1:  // START
 2:  M = 0 // flag to indicate the correction is done completely
 3:  SY = H . CDRi
 4:  if SY = 0
 5:      // No Errors in that code word
 6:      CDTi = CDRi
 7:  end if
 8:  Goto 31
 9:      if SY ≠ 0
10:      // Some Errors in that code word
11:      The type of bit error (single or double) and the EP value that will
        use to correct is specified depending on the SY value.
12:      Obtain SY value
13:      end if
14:      if SY value indicates there is a single bit error then
15:      // Obtain the EP from Lookup Table_1* that stored in memory
16:      CDTi = CDRi + EP // correction process
17:      else if SY value indicates there is a two bit error then
18:      // obtain the EP from Lookup Tabel_2* that stored in memory
19:      CDTi = CDRi + EP // correction process
20:      end if
21:      // after correction process, check the SY to make sure the
        correction is done completely
22:      SY = H . CDTi
23:      if SY = 0
24:      // The correction is done completely
25:      M=0
26:      return
27:      else
28:      // SY ≠ 0, the correction is NOT done completely, there are
            more than double bit error
29:      M=1 // flag to indicate there are more than double bit error
30.      end if
31.      return
```

Lookup Table_1*, include EP and SY for single bit error.
Lookup Table_2*, include EP and SY for double bit error.

Figure 10.3 Pseudocode for error detection and correction function of LCPC codes.

After the correction process is completed, SD_i can be obtained by implementing the decoding process on the $\overline{CD_{Ti}}$. The third process is the decoder, which is used to decode the $\overline{CD_{Ti}}$ and obtain the original source of the symbol data sent. The masking process for the last left four bits of the correct received code word must be done by AND operation for the $\overline{CD_{Ti}}$ with (111100000) as shown in Eq. (10.18).

$$SD_i = AND(\overline{CD_{Ti}}, 111100000) \tag{10.18}$$

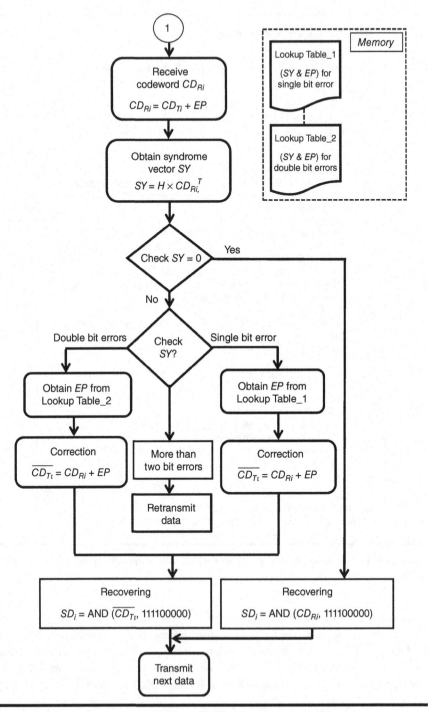

Figure 10.4   Flowchart of error detection, correction, and recovery for LCPC codes.

10.3 Complexity Analysis

This section presents the methodology of computational complexity analysis measurement of the LCPC codes. In order to measure the approach complexity, counts of additions and multiplications, in addition to the required memory size are used. Table 10.7 shows the memory size requirement of the three proposed types of LCPC codes, where n is the code word length, k is the symbol data length, and m (i.e., $n - k$) is the parity check bits length. NoEP is the number of error pattern for single- and double-bit error. NoSY is the number of syndrome vectors for single- and double-bit errors. The NoEP is computed based on Eq. (10.16). We compute the NoSY based on the simulation program by using the Eq. (10.11).

Table 10.8 shows the number of addition and multiplication operations for encoding and decoding processes of LCPC (9, 4) code. It also shows the Big O complexity. Table 10.9 shows the comparison complexity between the proposed LCPC (9, 4) code with different types of decoding algorithm for LDPC codes. Table 10.9 shows that the proposed LCPC (9, 4) code has low complexity $O(n)$ compared with the LDPC codes.

In addition, Table 10.10 illustrates the summary of comparison between the LDPC code and the proposed LCPC code in different parameters. Many differences are tabulated. Three main differences between the LDPC code and the proposed LCPC code are used for comparisons in this section. The first discrimination is that the complexity of the proposed LCPC code is lower than LDPC code. The second difference is the number of iterations in the proposed LCPC code. The third difference is the long latency of LDPC code compared with the proposed LCPC code in the decoding process that result from the longer code word length and number of iterations. These differences make the proposed LCPC code very useful in real-time communications for the futuristic wireless networks applications, due to the lower complexity and latency.

Table 10.7 Memory Size Required for Different Types of LCPC Codes

Type of code	Size of **G**	Size of **H**	Size of SY table	Size of EP table	Total size
LCPC (n, k)	matrix (n × k)	matrix n × (n − k)	(n-k) × NoSY	n × NoEP	(bits)
LCPC (9, 4)	9 × 4 = 36	9 × 5 = 45	5 × 30 = 150	9 × 30 = 270	501
LCPC (8, 3)	8 × 3 = 24	8 × 5 = 40	5 × 28 = 140	8 × 28 = 224	428
LCPC (7, 3)	7 × 3 = 21	7 × 4 = 28	4 × 23 = 92	7 × 23 = 161	302

Table 10.8 Number of Addition and Multiplication Operations for Encoding and Decoding Stages of LCPC (9, 4) Code

Operation Type	Encoding	Detection	Correction	Total No. of Operation	Error States
No. of MUL (n)	4	-	-	4	No Error
No. of MUL (m)	-	9	-	9	
No. of ADD (n)	3	-	-	3	
No. of ADD (m)	-	8	-	8	
Complexity	$O(4n + 3n)$	$O(9m + 8m)$	-	$O(4n + 3n + 9m + 8m)$	$O(7n + 17m)$ $= O(n)$
No. of MUL (n)	4	-	-	4	Single bit error or
No. of MUL (m)	-	9	-	9	Double bit errors with
No. of ADD (n)	3	-	1	4	one EP
No. of ADD (m)	-	8	-	8	
Complexity	$O(4n + 3n)$	$O(9m + 8m)$	$O(n)$	$O(4n + 4n + 9m + 8m)$	$O(8n + 17m)$ $= O(n)$

Table 10.9 Comparison Complexity of LDPC and LCPC Codes

Reference	Decoder Complexity	Note
(Davey and MacKay, 1998) [26]	$O(q^2)$	GF(q), $q = 64, 256$
(Barnault and Declercq, 2003)[27], (Chun-Hao et al., 2008) [28]	$O(q \log q)$	GF(q), $q = 64, 256$
(Declercq and Fossorier, 2007) [29]	$O(n_m q)$	$n_m < q$, $q = 64$, $n_m = 16$ or 32
(Voicila et al., 2010) [21]	$O(n_m \log n_m)$	$n_m < q$, $q = 64$, $n_m = 16$ or 32
(Xinmiao and Fang, 2011) [30]	$O(q^2 d_c)$	d_c is the check node degree for NB codes over GF(q)
(Proposed LCPC (9, 4) code)	$O(n)$	n is the code word length

Table 10.10 Comparison between LDPC code and the Proposed LCPC Code

Parameters	LDPC code	Proposed LCPC code
Code word length	Long	Short
Size of G matrix	Large	Small
Size of H matrix	Large	Small
Memory size requirement	Large	Small
Number of iteration	(10–50) time	0
Complexity of decoding	High ($O(q^2)$)	Low ($O(n)$)
Performance in AWGN channel	Good	Good
Performance in Fading channel	Poor	Good
Capability of burst error correction	No	Yes
Latency of decoding	Long	Short

10.4 Simulation Results

In this section, simulation results of LCPC codes are presented. The simulations are carried out to validate the performance of the proposed LCPC codes using binary phase shift keying (BPSK) modulation over an additive white Gaussian noise (AWGN) and Rayleigh fading channels. The BER performance of different code rates is investigated. Simulation results as shown in Figure 10.5 demonstrated the LCPC (7, 3) code has good BER performance compared with other codes over AWGN channel using BPSK modulation. Figure 10.5 shows that the LCPC (9, 4) code provides BER = 10^{-5} at SNR 7.3 dB, 7.1 dB in case LCPC (8, 3), and 6.9 dB in case LCPC (7, 3). The opportunity of bit error decreases for the short code word length. This explains the difference between the proposed LCPC codes.

The performance of the LCPC codes is compared with other codes, such as Hamming, BCH, RS, and LDPC codes [31] using various values of code word length. The Hamming, RS, BCH Soft, BCH Hard codes, and some decoding algorithms of the LDPC (8, 4) code such as Bit Flip, Log Domain and Log Domain Simple are implemented using MATLAB.

Figure 10.6 presents the comparison between LCPC, Hamming, RS, and BCH codes at code word lengths (7, 4), over AWGN channels using BPSK modulation. Figure 10.6 shows that the LCPC code improves the BER performance when compared with the Hamming, RS, and BCH codes. In LCPC (9, 4) code, the number of redundant parity bits is 5, so the maximum number of syndrome vectors obtained is 32. The LCPC (9, 4) code can detect and correct single-bit and many

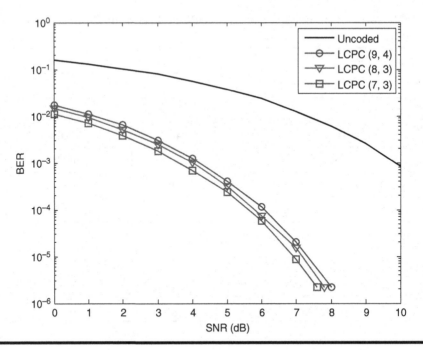

Figure 10.5   BER versus SNR for LCPC codes under AWGN and BPSK.

cases of double-bit errors as shown formerly in Tables 10.1 and 10. 2. In order to enable the LCPC code to correct all bit errors, we need to increase the number of the parity bits. This leads to increment in the code word length, and thus, increment in the possible number of error patterns. This can lead to a significant increment in the complexity of the error correction code.

Figure 10.6 shows that the LCPC (9, 4) code provides BER = 10^{-5} at SNR 7.3 dB, whereas the RS (7, 4) code provides the same BER (10^{-5}) in 9.4 dB SNR. The Hamming (7, 4) code provides the same BER (10^{-5}) in 9 dB SNR. Whereas, to obtain the same BER (10^{-5}), we need SNR equal to 8.1 dB for BCH Soft (7, 4) code, and 9.2 dB for BCH Hard (7, 4) code. The code gain that was obtained in this case is 2.1 dB over the RS (7, 4), 1.7 dB over the Hamming (7, 4), 0.8 dB over BCH soft (7, 4), and 1.9 dB over BCH Hard (7, 4) codes. The LCPC codes have the capability to correct single-bit and many cases of double-bit errors without needing iteration in the decoding process, whereas the Hamming (7, 4) code has the capability to correct one-bit error and RS (7, 4) code has the capability to correct one-bit error in each symbol.

The capability of the proposed LCPC code in error detection and correction is also studied. The minimum Hamming distance is defined as $d_{min} = n - k$, where $m \geq 3$ is a positive integer. The number of errors that a block code can detect and correct

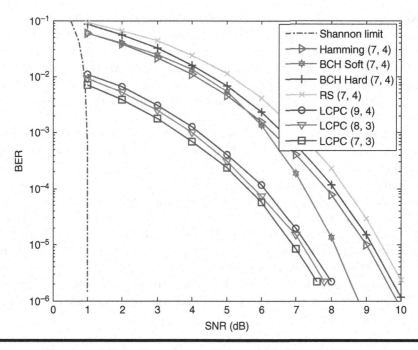

Figure 10.6 Comparison between LCPC (9, 4) code and other codes with Shannon limit over AWGN channel.

is determined by its minimum Hamming distance d_{min}. This is defined as the minimum number of places where any two code words differ. In general, the number of errors (u) that can be detected for a block code is $u = d - 1$. For example, at $m = 3$, the code word length $n = 7$, message length $k = 4$ and $d_{min} = 3$, where t is the number of errors that a block code can correct $t = [(n - k)/2]$. Since the Hamming code has a minimum Hamming distance $d_{min} = 3$, it can only correct 1 bit error for each 7 bits transmitted. Therefore, the error correction is $1/7 = 14.285\%$.

Likewise, in the case of RS codes, the number and type of errors that can be corrected depends on the characteristics of the RS code. The RS code is specified as RS (n, k) with s-bit symbols. This mean that the encoder takes k data symbols of s bits and adds parity symbols to make an $n = 2^s - 1$ symbol code word. There are $n - k$ parity symbols of each s bit. The RS decoder can correct up to t symbols that contain errors in an error code word, where $2t = n - k$. If $s = 3$ bits, $n = 7$, and when the number of parity is 3, $k = 4$. The number of symbols containing errors that RS code can correct is t, where $t = [(n - k)/2]$.

So, based on t value, the RS (7, 4) code can only correct one symbol error from the 7 code word symbols that are sent. If the symbol size is 3 bits, the worst case happens only when a one-bit error occurs in separate symbols. In this case, the error correction is $1/21 = 4.7619\%$, which is small compared with the percentage of error correction in the Hamming (7, 4) code, and this explains the reason why the Hamming (7, 4) code has a better BER performance when compared with the RS (7, 4) code, as shown in Figure 10.6. The best case for RS (7, 4) code occurs when all bits in a single symbol are wrong (or errors). This means that, the percentage of error correction is the number of errors in one symbol over the total number of symbols bit transmitted, i.e., $(3/21 = 14.285\%)$.

The saving power is one of the benefits of LCPC codes. Therefore, the proposed codes can be effectively used in WSN because of its huge reduction in the power consumption.

To investigate the BER performance of LCPC codes with the other codes that have code word length greater than the code word length of the LCPC codes, Figure 10.7 shows the comparison of BER performance of the proposed LCPC (9, 4) and (8, 3) codes with Hamming (15, 11) and RS (15, 11) codes. Figure 10.7 shows that the performance of LCPC codes is still better than the Hamming and RS codes, although the block length increases.

Similarly, Figure 10.8 presents the BER performance comparison between LCPC (9, 4) code and binary LDPC (8, 4) code with various decoding algorithms

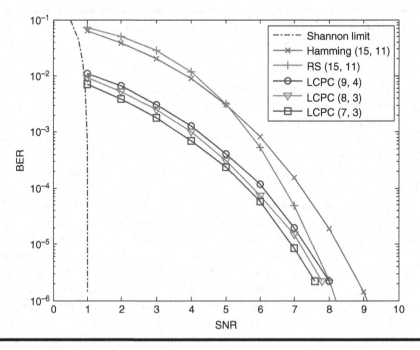

Figure 10.7   LCPC codes versus the Hamming (15, 11) and RS (15, 11) codes over AWGN channel.

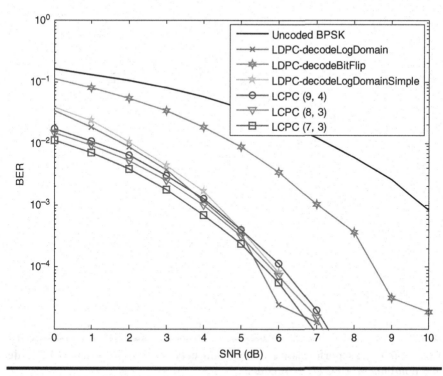

Figure 10.8   Comparison between LCPC (9, 4) code and binary LDPC (8, 4) with different types of decoding using BPSK over AWGN channel.

using BPSK modulation over AWGN channel. The binary LDPC (8, 4) code is implemented for three different types of decoding algorithms (i.e., Bit Flip, Log Domain, and Log Domain Simple). The performance of LCPC (9, 4) code is better than the binary LDPC code when the Bit Flip decoding method (BF) is used, and the coding gain is around 3 dB at BER = 10^{-5}, whereas, the coding gain is 1 dB at BER = 10^{-5} in case Log Domain decoding algorithms are used. The error correction capability of binary LDPC code depends on the code word length and the characteristic of the parity check matrix. The decoder gives a better performance with a larger code word (i.e., big size of **G** and **H** matrices). The matrix multiplication for that larger code word length demands large memory size, computational requirements and more complex decoding [5, 7].

In addition to the better performance of the BER, the main advantage of the proposed LCPC code is the low complexity of encoding and decoding process when compared with the LDPC and RS codes. Furthermore, the proposed code does not require reiteration during the decoding of error correction, which is a salient feature of this code and is an important improvement over previous codes.

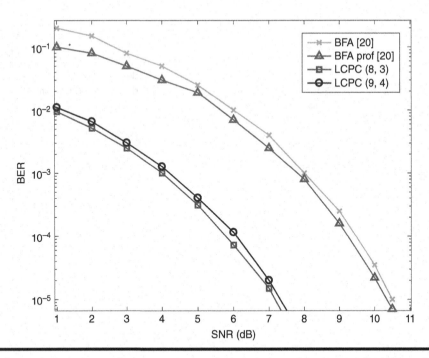

Figure 10.9 BER performance comparison between the proposed LCPC code and non-binary LDPC (7; 3; 4) code.

Figure 10.9 shows the comparison of BER performance of proposed LCPC (9, 4) and (8, 3) codes with non-binary LDPC (7; 3; 4) code in Bit-Flipping A (BFA) algorithm and Bit-Flipping A probabilistic (BFA prob) algorithm is used [32]. We assume block length ($n = 7$), column weight ($c_w = 3$) and row weight ($r_w = 4$). Figure 10.9 shows that the LCPC (9, 4) code provides BER = 10^{-5} at SNR 7.3 dB, and 7.1 dB in case LCPC (8, 3) is used. However, the LDPC code provides the same BER (10^{-5}) after experiencing the 10.5 dB SNR. As for the obtained code gain, it is 3.2 dB and 3.4 dB for LCPC (9, 4) and LCPC (8, 3) codes, respectively.

Figure 10.10 shows the BER performance comparison between the LCPC (9, 4) and (8, 3) codes with non-binary LDPC codes, lower diagonal-based PCM (LDM) and doubly diagonal- based PCM (DDM) for GF(4) [33]. Figure 10.10 demonstrates that the BER performance of proposed LCPC codes is better than the codes that presented in [33]. Figure 10.10 shows that the LCPC (9, 4) code provides BER = 10^{-5} at SNR 7.3 dB, and 7.1 dB in case LCPC (8, 3) code, whereas the non-binary LDPC code in case GF(4) with DDM provides the BER (0.5×10^{-3}) in 7 dB SNR. The non-binary LDPC code in case LDM and PCM provides the BER (0.2×10^{-2}) and (10^{-2}) in 7 dB SNR, respectively [33].

The BER performance of the proposed LCPC codes over the Rayleigh fading channel using BPSK modulation is investigated. Figure 10.11 shows the BER

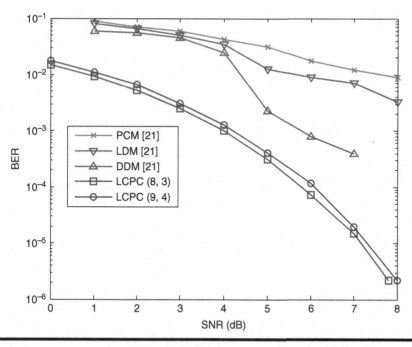

Figure 10.10   BER performance comparison between proposed LCPC codes and PCM, LDM and DDM for non-binary LDPC code GF(4) based on FFT-SPA decoding.

performance comparison between the LCPC (9, 4) code and LDPC (576, 288) with 41472 bits. Figure 10.11 shows that the LCPC (9, 4) code outperforms the LDPC (576, 288) code over Rayleigh fading channel.

10.5 Conclusions

In this chapter, a short code word-length approach has been proposed. The performance of the proposed LCPC codes with BPSK modulation over AWGN and Rayleigh fading channels is investigated. The BER performance comparisons are made between the LCPC codes with Hamming, RS, BCH, binary and non-binary LDPC codes. The simulation results show significant enhancement in the BER performance of LCPC code as compared to the renowned LDPC, RS, BCH, and Hamming codes. The LCPC code has characteristics that distinguish it from LDPC codes, such as low complexity in the encoding and decoding processes, low memory size requirement (only 501 bits for LCPC (9, 4)), and no iterations in decoder process when compared with LDPC codes that need more than 20 times of iterations to correct the error code word.

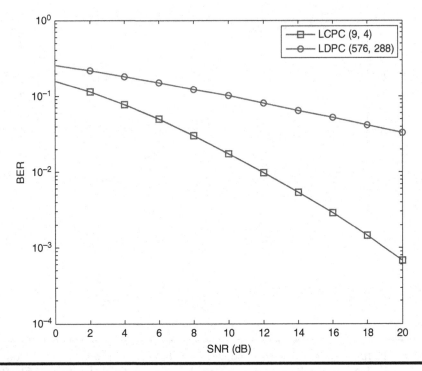

Figure 10.11   Comparison between LCPC (9, 4) code and LDPC (576, 288) at BPSK over Rayleigh fading channel at 41472 bit.

References

1. Kaythry P., Kishore R., Praveena V. 2018. Energy efficient raptor codes for error control in wireless body area networks. *Wireless Personal Communications*, 1–19, DOI. 10.1007/s11277-018-5271-y.
2. Al-Turjman F. 2017. Cognitive caching for the future fog networking. *Elsevier Pervasive and Mobile Computing*, DOI. 10.1016/j.pmcj.2017.06.004.
3. Al-Turjman F. 2016. Impact of user's habits on smartphones' sensors: An overview. In *HONET-ICT International IEEE Symposium*, Kyrenia, Cyprus, 70–74.
4. Al-Turjman F. 2016. Hybrid approach for mobile couriers election in smart-cities. In *Proc. of the IEEE Local Computer Networks* (LCN), Dubai, UAE, 507–510.
5. Tsimbalo E., Fafoutis X., Piechocki R.J. 2017. RC Error correction in IoT applications. *IEEE Transactions on Industrial Informatics*, 13(1), 361–369.
6. Huang J., Zhou S., Willett P. 2009. Near-Shannon limit linear-time-encodable nonbinary irregular LDPC codes. In *IEEE Global Telecommunications Conference*, GLOBECOM 2009, 1–6.
7. Zhong H., Zhang T. 2005. Block-LDPC: A practical LDPC coding system design approach. *IEEE Transactions on Circuits and Systems I: Regular Papers*, 52, 766–775.

8. Cole C.A., Wilson S.G., Hall E.K., Giallorenzi T.R. 2006. Regular 4, 8 LDPC codes and their low error floors. In *IEEE Military Communications Conference*, MILCOM 2006, 1–7.

9. Carrasco R.A., Johnston M. 2009. Non-binary error control coding for wireless communication and data storage. Hoboken, NJ: Wiley.

10. Alabady S., Al-Turjman F. 2019. A novel approach for error detection and correction for efficient energy in wireless networks. *Springer Multimedia Tools and Applications*, 78(2), 1345–1373.

11. Kou Y., Lin S., Fossorier M.P.C. 2001. Low-density parity-check codes based on finite geometries: A rediscovery and new results. *IEEE Transactions on Information Theory*, 47, 2711–2736.

12. Chen X., Men A. 2008. Reduced complexity and improved Performance Decoding Algorithm for Nonbinary LDPC Codes Over GF (q). In *11th IEEE International Conference on Communication Technology*, ICCT 2008, 406–409.

13. Wang C.L., Chen X.H., Li Z.W., Yang S.H. January 2013. A simplified min-sum decoding algorithm for non-binary LDPC codes. IEEE Transactions on Communications, 61(1), 24–32.

14. Richardson T.J., Shokrollahi M.A., Urbanke R.L. 2001. Design of capacity-approaching irregular low-density parity-check codes, *IEEE Transactions on Information Theory*, 47, 619–637.

15. Salbiyono A., Adiono T. 2010. LDPC decoder performance under different number of iterations in mobile wimax. In *International Symposium on Intelligent Signal Processing and Communication Systems* (ISPACS), Chengdu, China, 1-4.

16. Baldi M., Chiaraluce F., Maturo N., Liva G., Paolini E. 2014. A hybrid decoding scheme for short non-binary LDPC codes. *IEEE Communications Letters*, 18(12), 2093–2096.

17. Ranganathan S.V.S., Divsalar D., Wesel R.D. 2017. Design of improved quasi-cyclic protograph-based raptor-like LDPC codes for short block-lengths. In *IEEE International Symposium on Information Theory* (ISIT), Aachen, Germany, 1207–1211.

18. Nguyen T.V., Nosratinia A. 2013. Rate-compatible short-length protograph LDPC codes. *IEEE Communications Letters*, 13(5), 948–951.

19. Bocharova I.E., Hug F., Johannesson R., Kudryashov B.D. 2012. High-rate QC LDPC codes of short and moderate length with good girth profile. In *7th International Symposium on Turbo Codes and Iterative Information Processing* (ISTC), Gothenburg. Sweden, 150–154.

20. Alabady S., Al-Turjman F. 2018. LCPC error correction code for Internet of Things applications. *Elsevier Sustainable Cities and Society*, 42, 663–673.

21. Voicila A., Declercq D., Verdier F., Fossorier M., Urard P. 2010. Low-complexity decoding for non-binary LDPC codes on high order fields. *IEEE Transactions on Communications*, 58, 1365–1375.

22. Trifina L., Ryu J., Tarniceriu D. 2017. Up to five degree permutation polynomial interleavers for short length LTE turbo codes with optimum minimum distance. In *International Symposium on Signals, Circuits and Systems* (ISSCS), Iasi, Romania, 1–6.

23. Cojocariu E., Tarniceriu D., Trifina L., Lazar A.G. 2010. Performance of asymmetric turbo codes on Rayleigh fading channels for small interleaver length. *2010 3rd International Symposium on Electrical and Electronics Engineering* (ISEEE), Galati, Romania, 54–57.

24. Liva G., Paolini E., Matuz B., Scalise S., Chiani M. 2013. Short turbo codes over high order fields. *IEEE Transactions on Communications*, 61(6), 2201–2211.

25. Andrei M., Trifina L., Tarniceriu D. 2013. Influence of trellis termination methods on turbo code performances. 4th *International Symposium on Electrical and Electronics Engineering* (ISEEE), Galati, 1–6.

26. Davey M.C., MacKay D.J. 1998. Low density parity check codes over GF (q). *Information Theory Workshop*, 70–71.

27. Barnault L. Declercq D. 2003. Fast decoding algorithm for LDPC over GF (2q). In *the Proceeding Information Theory Workshop*, 70–73.

28. Chun-Hao L., Chien-Yi W., Chun-Hao L., Tzi-Dar C. 2008. An O(qlogq) log-domain decoder for non-binary LDPC over Gf(q). In *IEEE Asia Pacific Conference on Circuits and Systems*, APCCAS 2008, 1644–1647.

29. Declercq D., Fossorier M. 2007. Decoding algorithms for nonbinary LDPC codes over GF(q). *IEEE Transactions on Communications*, 55, 633–643.

30. Xinmiao Z., Fang C. 2011. Reduced-complexity decoder architecture for non-binary LDPC codes. *IEEE Transactions on Very Large Scale Integration (VLSI) Systems*, 19, 1229–1238.

31. Chen C.-Y., Huang Q., Chao C.-c., Lin S. 2010. Two low-complexity reliability-based message-passing algorithms for decoding non-binary LDPC codes. *IEEE Transactions on Communications*, 58, 3140–3147.

32. Miladinovic N., Fossorier M.P.C. 2005. Improved bit-flipping decoding of low-density parity-check codes. *IEEE Transactions on Information Theory*, 51, 1594–1606.

33. Aruna S., Anbuselv M. 2013. FFT-SPA Based non- binary LDPC decoder for IEEE 802.11n Standard. In *International Conference on Communications and Signal Processing (ICCSP)*, 566–569.

Chapter 11

Energy-Harvesting Methods for WBAN Applications

Süleyman Mahircan Demir[1],
Fadi Al-Turjman[2], and Ali Muhtaroğlu[3]

[1]*Department of Electrical and Electronics Engineering, Middle East Technical University Northern Cyprus Campus, Kalkanli, Güzelyurt, Turkey*
[2]*Department of Computer Engineering, Antalya Bilim University, Antalya, Turkey*
[3]*Center for Sustainability, Middle East Technical University Northern Cyprus Campus, Kalkanli, Güzelyurt, Turkey*

Contents

11.1 Introduction 204
11.2 Microscale Energy Scavenging Classification 206
    11.2.1 Photovoltaic Energy Scavenging 207
    11.2.2 Thermoelectric Energy Scavenging 208
    11.2.3 Vibration-Based Energy Scavenging 210
    11.2.4 Radio Frequency (RF) Energy Scavenging 212
    11.2.5 Hybrid Energy Scavenging 213
11.3 Wireless Body Area Networks 214
11.4 Energy Scavenging for WBANs 217
    11.4.1 Photovoltaic Energy Scavenging for WBANs 217
    11.4.2 Thermoelectric Energy Scavenging for WBANs 219

11.4.3 Piezoelectric Energy Scavenging for WBANs 222
11.4.4 RF Energy Scavenging for WBANs..223
11.4.5 Hybrid Energy Scavenging for WBANs...224
11.5 Conclusion and Open Issues 226
References 228

11.1 Introduction

Demand for technology has tremendously increased over the last few decades. This demand has been accompanied by industrial developments. Therefore, energy needs for modern society has become critically important. At this point the energy scavenging, i.e., energy harvesting, concept has started to attract considerable attention to solve energy problems and to save existing energy resources for the next generation. Essentially, energy scavenging is utilization of the ambient energy sources to generate electrical energy. For this reason, the concept of energy scavenging is not only important for extracting energy from the ambient sources but also for sustainability. Thus, solving energy problems with a sustainable method is very interesting for many professionals working on energy-scavenging technologies.

Internet of Things (IoT) is a relatively new term which refers to interconnection of physical devices from microscale electronic circuits to large servers through the internet [1, 2]. A white paper by Cisco reported that total number of connected devices in 2010 was 12.5 billion. It is expected that this number will be around 50 million by the year 2020. Also, the number of connected devices per capita is predicted to increase from 1.84 to 6.58 for the same period [3]. These connected devices, which can store, process, and send/receive data, change our lifestyle and help us in our daily life. Wireless sensor networks (WSNs) are important examples of the IoT concept. WSNs are wireless network systems consisting of sensors and relay nodes which can have different power consumption, data-handling capacity, and communication range [4]. These networks are mainly used for monitoring physical and environmental conditions, positioning and animal tracking, security and surveillance, logistics, entertainment, transportation, and industrial applications [5].

Lately, research has focused on utilization of WSNs for healthcare systems. These systems are generally referred as wireless body area networks (WBANs). There are millions of people who suffer from chronic or fatal diseases. Many studies show that these diseases can be prevented when they are detected in early stages. In other words, early diagnosis is vitally important for the treatment of the diseases [6]. Therefore, utilization of WBANs to monitor abnormal conditions happening in the human body substantially increases the quality of life.

Typically, batteries are used to supply power to WBANs. However, finite capacity and replacement necessity of batteries are significant problems, especially for biomedical applications [7]. Increasing the battery size is not an optimal solution, as the cost and weight simultaneously increase [8]. Increase in weight creates a bulky system and hurts the mobility. Therefore, recent studies focus on removing batteries from WBANs.

Microscale energy scavenging is one of the most popular methods to increase battery life or completely remove the battery from ultra-low-power electronic systems [1, 9]. Removing batteries from the system would be the ideal case, particularly for the sensor networks which are used for biomedical applications [10]. However, it is essential to identify the availability of sources for energy-scavenging circuits to evaluate whether the corresponding scavenging technology is useful for a specific application. In general, solar, thermal, mechanical movement or vibration, and ambient radio-frequency (RF) sources can be used to generate electrical energy [11]. The vibration or movement-based energy scavenging can be divided into three major categories: electrostatic, electromagnetic, and piezoelectric [12–14] (Figure 11.1). Thus, the availability of one of these sources is one of the most crucial criteria for a specific application. Furthermore, energy extraction capacity, physical size, robustness, and output impedance characteristics of the harvesting circuits play an important role for the potential applications.

In this chapter, microscale energy-scavenging methods and leading harvesting architectures are carefully investigated, particularly for WBAN applications. There are numerous studies in the literature on energy scavenging to extract maximum energy with the best efficiency. Some of the proposed architectures and their working principles are summarized to provide a general insight into the harvesting methods. Harvesting architectures are classified by considering the source of energy which they utilize to generate electricity. Moreover, the output powers, sizes, and conversion efficiencies are provided to simplify the comparison of the structures.

Most of the existing surveys on energy scavenging deal with the classification of energy-scavenging methods based on their source types. The study presented in [15] focuses on energy-scavenging methods, and it briefly explains the working principles of each method. In [16], scavenging methods are classified by considering

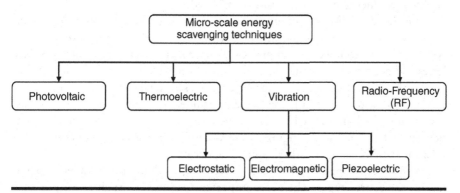

Figure 11.1   Microscale energy scavenging techniques [14].

WSN applications. Similarly, [17] presents the classification of the scavenging methods specifically for WSNs. Another study presented in [18] introduces energy-scavenging methods for WBANs. However, the study briefly mentions the scavenging methods and some of the work found in the literature to provide a general perspective. In this chapter, energy-scavenging methods and some of the interface circuits, particularly for WBANs, are presented. The interface circuits summarized in this survey article have been proposed particularly for WBAN applications. Moreover, the detailed analyses of power management units for each interface circuit are also provided. Furthermore, some of the presented works in this paper have a prototype which increases the possibility of the real-life integration.

The rest of the chapter is organized as follows: Section 11.2 provides background information about energy-scavenging techniques and classification of the scavenging methods based on the ambient energy sources. In this section, general characteristics and basic principles of different scavenging methods are specified. Section 11.3 talks about WBANs which are examined for any possible integration with the harvesting systems. In Section 11.4, harvesting architectures specifically for WBANs are investigated to analyze whether the corresponding architectures are suitable for WBANs in terms of output power, physical size, and environmental conditions. In this section, pros and cons of each architecture are discussed in detail. Finally, Section 11.5 summarizes the study and remarks on future work.

11.2 Microscale Energy Scavenging Classification

The aim of energy scavenging is to extend the battery life or replace the battery. However, it is not an easy process to have an efficient scavenging system with high output power which is high enough to energize a sensor network. In fact, there are many parameters to check whether the corresponding energy-scavenging system is suitable for a specific WBAN application. Availability of the ambient source, output power, output impedance, and physical size of the system are some of the parameters which should be carefully considered.

In realistic conditions, output power of a microscale energy-scavenging circuit varies between microwatts and milliwatts, which is relatively low power to run a system on. Also, complicated systems which have many interface circuits introduce losses and consequently reduce the efficiency of the circuit. Therefore, the integration of these interface circuits in order to increase the efficiency via reducing the cost and size is proposed in [1]. However, this integration introduces size and cost constraints which make the life more difficult for a circuit designer. Thus, integration level of an energy-scavenging circuit is another important parameter to consider for WBANs and any other potential application.

In this section, microscale energy-scavenging methods are classified depending on the source types. The other important parameters, such as output voltage and power levels, output impedances, and physical sizes are also briefly

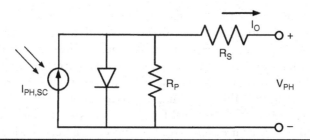

Figure 11.2　Electrical model of PV panel [21].

mentioned. At the end of the section, general characteristics of the scavenging methods are presented in Table 11.3 to compare the advantages and disadvantages of each method.

11.2.1 Photovoltaic Energy Scavenging

Photovoltaic (PV) energy-scavenging devices use either sunlight or any artificial light to produce electricity. Basically, PV cells which are composed of semiconductor materials absorb the light from outside source. Thanks to the p-n junction effect, absorbed light releases electrons from the semiconductor. The released electrons and holes are collected at the electrodes to create a voltage difference [19, 20].

Figure 11.2 shows the electrical model of a PV cell [14, 21]. $I_{PH,SC}$ is the generated photocurrent. A forward-biased diode and equivalent shunt resistance R_P are shown parallel to the photocurrent. Serial resistance R_S represents the parasitic resistance. The current I_0 and the voltage V_{PH} are the output current and voltage of the PV cell. As the output voltage of a PV cell is in the form of direct current (DC), the output does not need a rectification and can be directly stored or used for electronic devices depending on the voltage level. A generic model of a PV harvesting system presented by [22] is shown in Figure 11.3.

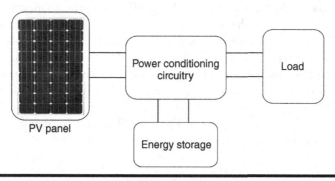

Figure 11.3　Photovoltaic energy scavenging system [22].

Primary concern about PV energy scavenging is the alteration in temperature or irradiance. The output characteristic of a PV cell changes non-linearly with the varying temperature or irradiance [23]. Since the behavior is non-linear, it becomes very hard to detect the optimal voltage and current levels at which the harvester should function to obtain the maximum power. Therefore, maximum power point tracking (MPPT) techniques are required for PV energy scavenging. There are plenty of studies on MPPT structures in the literature [24–27]. In general, MPPT structures can be divided into two groups: with digital signal processor (DSP) or microcontroller, and without DSP or microcontroller. DSPs and microcontrollers are mainly used for macroscale PV power systems, while they are not desirable for microscale PV systems, since they introduce additional power consumption [23].

Even though different MPPT techniques are applied to PV energy-scavenging circuits, environmental conditions affect this kind of harvesting profoundly. Therefore, indoor and outdoor power densities are significantly different. Typically, PV energy scavenging provides a power density of 100 μW/cm^2 to 1000 μW/cm^2 indoors and 100 mW/cm^2 outdoors with the efficiency of 30% for monocrystalline cells [28]. However, the efficiency of 30% does not reflect the actual potential of PV cells. Conventional PV cells usually absorb light in the visible light frequencies range, and studies focus on increasing the efficiency by using nanotechnology and harvesting energy in UV spectrum range frequencies [29]. Despite high dependence on the lighted environment, PV energy-scavenging systems are still very attractive for many applications due to their well-developed structure and high power densities.

11.2.2 Thermoelectric Energy Scavenging

In order to extract energy from temperature difference, thermoelectric generators (TEGs) are employed. In principle, TEGs are composed of thermocouples which contain p- and n-type semiconductors. They are connected electrically in series and thermally in parallel [30]. There are three main effects for the operation of TEGs: the Seebeck effect, the Peltier effect, and the Thomson effect. The most interesting one for the energy scavenging is the Seebeck effect due to the fact that it claims an electromotive force can be produced when the junction between two different materials preserved at different temperatures [19]. In other words, the generator produces electrical energy by using the temperature gradient between hot and cold surfaces. Figure 11.4. shows the basic operation of a TEG.

The maximum efficiency of a TEG is calculated by using the Carnot efficiency [28]:

$$\eta = \frac{(T_H - T_C)}{T_H} \qquad (11.1)$$

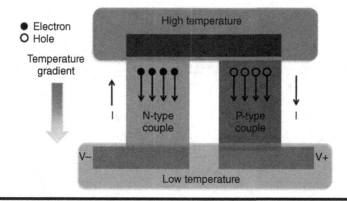

Figure 11.4   Basic operation of a TEG [14].

T_H and T_C in Eq. (11.1) represent the hot and cold temperatures, respectively, in Kelvin. An equivalent electrical circuit model of a TEG presented by [31] is shown in Figure 11.5. In order to show the effect of the temperature difference to the output current, voltage, and power, simulation results for Micropelt [32] MPG-D751 commercial miniature TEG are presented by [21], using the simulation tool provided by the manufacturer [33]. Figure 11.6 shows the corresponding simulation results.

As it can be realized from Figure 11.6, the output voltage of the TEG is between 0–0.6V and the maximum power is around 250 μW. Increase in temperature gradient clearly increases the output voltage. The output voltage and power of the TEG are lower compared to the PV energy harvesters, especially for outdoor conditions. However, TEGs are less intermittent since they work as long as there is a temperature difference, ΔT, across their surfaces. Furthermore, their highly reliable, easily scalable, and stationary structure, together with relatively long life, make them very attractive principally for body area network (BAN) applications, since human body temperature can be easily utilized to extract power from TEGs [15, 17, 34]. Also, the output of a TEG is also in the form of direct-current (DC) like the PV energy harvesters, and it does not require a rectification. However, TEGs may need an interface circuit to boost the output voltage in order to make the voltage level high enough to drive a sensor node [31, 34].

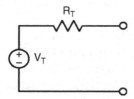

Figure 11.5   Equivalent electrical circuit model of a TEG [31].

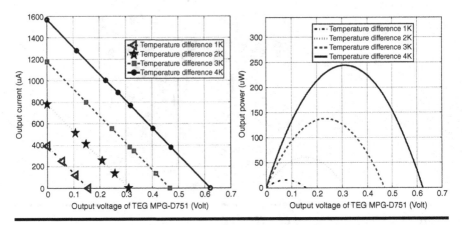

Figure 11.6   Simulation results for MPG-D751 commercial TEG [33].

11.2.3 Vibration-Based Energy Scavenging

Alternatively, vibration or mechanical movement can be also used as a source for energy harvesting. As depicted in Figure 11.1, there are three different mechanisms of converting vibrations into electrical energy. These mechanisms are: electromagnetic, electrostatic, and piezoelectric.

Energy conversion principle for electromagnetic mechanism is based on Faraday's law of induction. A generator induces an alternating current in a coil when there is a relative motion between the coil and a magnetic field [14]. The reason why this type of conversion mechanism is considered as vibration- or mechanical movement-based is that the relative motion between the coil and a magnet can be created by using vibration or mechanical movement [35]. However, huge coil and magnet requirements of the system make small-scale implementation of this type of conversion mechanism very hard, especially for high-power applications [14, 36].

Electrostatic conversion mechanism depends on a microelectromechanical systems (MEMS) variable capacitor. Initially charged capacitor is placed in a system such that a mechanical movement changes the positions of the conductors which forms the capacitor. Variation in the positions of the conductors changes the capacitance value. Therefore, the energy stored in the capacitor also changes, and mechanical energy is converted into electrical energy which can be stored in a battery or used by an external load.

However, since the capacitor should be initially charged, a separate voltage source is necessary for this type of mechanism [13, 35, 37].

The principle of the piezoelectric mechanism is based on charge separation. The mechanical strain in a piezoelectric material causes charge separation across the dielectric material. This charge separation creates a potential difference. The advantage of this method is that piezoelectric transducers do not require a separate

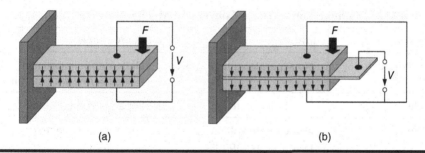

(a)   (b)

Figure 11.7   Piezoelectric bimorph configuration for (a) series and (b) parallel operation [38].

voltage source [13, 14]. For piezoelectric energy harvesting, the amount of mechanical energy which is converted into electrical energy is measured by a parameter which is called a coupling factor. A high coupling factor indicates that amount of mechanical energy converted into electrical energy is large [38].

Generally, piezoelectric bimorphs are used for energy harvesting applications. Figures 11.7 and 11.8 show two different piezoelectric bimorph configurations for series and parallel operations. As it can be realized from the figure, there are two piezoelectric layers attached together for both configurations. The bimorph configuration depicted in Figure 11.7(a) is for series operation, since two piezoelectric layers are poled in the opposite direction. Hence, the voltage is doubled while the piezoelectric capacitance becomes half of its original value and the current stays the same in comparison to the one-layer case. On the other hand, the bimorph configuration in Figure 11.7(b) shows the parallel connection. For this type of configuration, the current and the piezoelectric capacitance are doubled while the voltage level is the same. However, it should be noted that different piezoelectric configurations only change the voltage and current ratio. The amount of deliverable output power is not affected by the type of layer connections [38].

In this chapter, applications of piezoelectric transducers (PZT) are investigated in detail due to their high potential and easy utilization for biomedical applications and WBANs [1, 39]. However, in order to compare these three mechanisms,

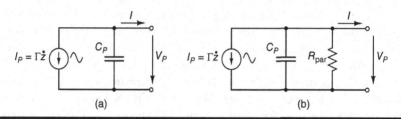

(a)   (b)

Figure 11.8   Equivalent electrical model of a PZT (a) neglecting and (b) considering the dielectric losses [38].

Table 11.1 Comparison of Three Conversion Mechanisms Based on [1, 13]

Mechanism	Advantages	Disadvantages
Piezoelectric	- No separate voltage source	- High output impedance - Complex interface circuits
Electrostatic	- Easy to integrate	- Separate voltage source
Electromagnetic	- No separate voltage source - Low output impedance - Simple interface circuits	- Low output voltage for small circuits

Table 11.1 is presented. Table 11.1, based on [1, 13], briefly presents the advantages and disadvantages of these three different conversion mechanisms to provide a general perspective.

11.2.4 Radio Frequency (RF) Energy Scavenging

RF signals are constantly available in the air, particularly in big cities where the population density is high. Therefore, harvesting RF signals from the environment enables wireless charging of WBANs [40]. Also, since RF scavenging is not affected by environmental conditions like temperature and weather, it becomes very attractive for WBAN applications. Ambient RF sources can be listed as [41]:

- AM radio band (550 kHz–1605 kHz)
- FM radio band (87.5 MHz–108 MHz)
- TV band (41MHz– 950 MHz)
- GSM band (0.85–0.90 GHz, 1.8–1.9 GHz)
- CDMA band
- 3G band
- 4G band
- ISM band (2.4 GHz)
- Wi-Fi band (2.45 GHz–5.8 GHz)

Figure 11.9 represents the RF energy-scavenging system. The procedure for RF energy scavenging starts with capturing radio signals from the ambient. An antenna is used for this capturing procedure. Impedance matching circuits are integrated into the system in order to match the rest of the circuit with the antenna so that a reflection does not occur. The captured RF signals are converted into DC by using different methods depending on applications. Afterward, the rectified signal is sent to the power management unit to be used in WBAN applications. The source power, antenna gain, distance between source and antenna, and energy conversion efficiency are some of the parameters which affect the amount of harvested power for RF energy-scavenging systems [17].

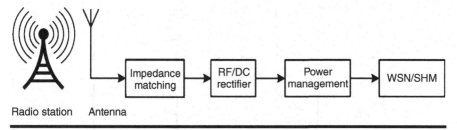

Radio station   Antenna

Figure 11.9   RF energy-scavenging system [41].

The output density of ambient RF scavenging is between $0.0002–1 \ \mu W/cm^2$ [1]. This output power range is relatively less compared to the other energy-scavenging methods. Another important point for RF scavenging is the distance sensitivity. Table 11.2 provides efficiencies, energy levels, and limitations for different RF sources. The limitation column shows the distance sensitivity depending on the RF source. However, the continuous availability of the RF signal for both indoor and outdoor conditions makes the RF scavenging method an attractive approach for WBANs applications. Also, the possibility of antenna integration and developing antenna design techniques increases the popularity of the RF scavenging method for the future applications.

11.2.5 Hybrid Energy Scavenging

Recently, studies on energy scavenging also focus on hybrid energy harvesting technologies. Fundamentally, hybrid energy harvesting is using multiple harvesting modes simultaneously to extract energy in order to meet the energy need of WBANs [1]. As hybrid harvesting systems can provide higher power, they can extend the on-board features of WBANs and can improve their communication range.

The total number of sources can be two or more for hybrid harvesters. Studies for TEG-PZT, PZT-RF, and TEG-RF combinations can be found in the literature. Also, a triple source hybrid scavenging system was recently presented by [42].

Table 11.2   RF Energy Scavenging Comparison for Different Frequencies [41]

Source	Efficiency	Energy Level	Limitation
RF – GSM	Low	mW	0–100 m
RF – TV	Low	µW	0–4 km
RF – WIFI	Low	nW - µW	0–10 m
RF – AM	Low	µW - mW	0–20 km

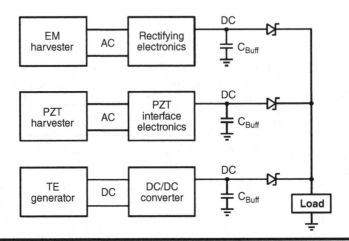

Figure 11.10 **Hybrid harvester structure with triple sources [42].**

The harvester structure presented in [42] is shown in Figure 11.10. The system combines TEG, vibration-based electromagnetic (EM), and PZT harvesters. Therefore, the proposed structure is able to extract energy simultaneously from three independent sources. Also, it provides a single DC output as it can be realized in the same figure. It is reported that simultaneous multi-mode operation of the harvester can generate up to 110 μW output power.

Since hybrid scavenging is relatively new field of study compared the other scavenging methods, it can be expected that different hybrid structures with higher efficiency and output power will be seen in the future. Furthermore, the studies are generally focused on combining two harvesting methods [1, 42]. This trend can change for combining all harvesting methods in one circuit to extract the maximum available energy from the environment to remove the battery from WBANs or increase its capability and communication range.

Table 11.3 compares the advantages and disadvantages of each method presented in this Section.

11.3 Wireless Body Area Networks

In order to utilize WSNs for healthcare monitoring systems without affecting human life, micro- and nanoscale sensor systems with very low power consumptions are needed. These systems generally referred as WBANs. Utilization of WBANs may save lives and significantly reduce the cost of in-hospital monitoring. It is possible to place WBANs on the human body or implant them into the body. Also, monitoring the human body using wearable WBAN systems is quite an attractive option, especially for medical applications [43]. WBANs can be used not only for medical applications but also for non-medical applications such as real-time streaming, entertainment, and non-medical emergency situations [6]. WBAN applications are presented in Figure 11.11.

Table 11.3 Energy Harvesting Methods and Their Characteristics [1]

	PV Solar	Thermoelectric	Piezoelectric vibration	Electromagnetic vibration	Ambient RF
Power density	Outdoor:100mW/cm^2 Indoor:<100μW/cm^2	50–100 μW/cm^2 per ^{0}C	10–200 μW/cm^3	1–2 μW/cm^3	0.0002–1 μW/cm^2
Output voltage	0.5 V max	10–100 mV	10–20 V (open ckt)	few 100mV	3–4 V (open ckt)
Availability condition	Lighted environment	Surfaces with ΔT	Hz–kHz Vibration	Hz Vibration	Vicinity to radiation source
Pros	High power density Well developed technology	Non-intermittent/ less intermittent than alternatives	High voltage Well developed technology	Well developed	Antenna can be integrated Widely available
Cons	Intermittent Highly dependent on light	Low voltage Need ΔT	Highly variable output Large area High output impedance	Bulky Low power density Low output voltage	Very sensitive to distance of the RF source

			Assessing soldier fatigue and battle readiness
WBAN Applications	Medical	Wearable WBAN	Aiding professional and amature sport training
			Sleep staging
			Asthma
			Wearable health monitoring
		Implant WBAN	Cardiovascular diseases
			Cancer detection
		Remote control of medical devices	Ambient Assisted Living (AAL)
			Patient monitoring
			Tele-medicine systems
	Non-medical		Real time streaming
			Entertainment applications
			Emergency (non-medical)

Figure 11.11   WBAN applications [6].

A WBAN is composed of sensor/actuator nodes. Each node has enough capability to perform its tasks. Moreover, each node has its own energy supply which contains a storage element and an energy harvesting device. For each node, it is possible to communicate with the other sensor nodes. Also, a central node can be employed to have communication with outside world by using a standard telecommunication infrastructure. Thus, it is possible to provide services such as chronic disease management, medical diagnostic, and fitness tracking to a WBAN user. Furthermore, WBAN is expected to provide feedback to the individual about her/his lifestyle and health status in the future [44]. A conceptual diagram of WBANs is shown in Figure 11.12.

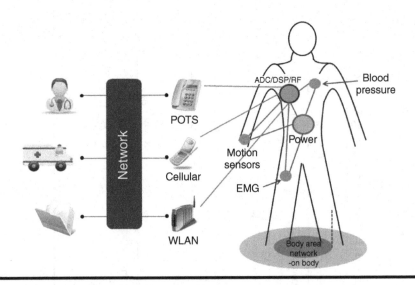

Figure 11.12   WBAN conceptual diagram [45].

11.4 Energy Scavenging for WBANs

In this section, different energy-harvesting interface circuits which are used to power WBANs are investigated. Classification of energy-scavenging methods based on their source types has been provided in Section 11.2. In order to compare different scavenging methods, this chapter focuses on energy-harvesting circuits specifically for WBANs.

11.4.1 Photovoltaic Energy Scavenging for WBANs

Detailed analysis of PV energy scavenging, together with its pros and cons, have been presented in Section 11.2. In summary, photovoltaic energy harvesting systems are well-developed technologies, and they have high-output power density. However, PV harvesting systems highly depend on temperature and irradiation. The sensitivity to outside conditions makes the implementation of PV energy scavenging more difficult for indoor conditions where the irradiation is limited.

In [46], a PV energy harvesting system is proposed for WBAN applications. The system is supplied by a thin-film PV harvester. The harvester ensures that the system is working properly even in indoor conditions. The proposed system is fully autonomous and does not need a battery even for the start-up. Even though there is no battery, the harvested energy is stored in a supercapacitor. Once there is enough energy, data is collected from the sensors and transmitted to the receiver. The frequency of the data transmission rate is given by the formula:

$$f = \frac{(P_H - P_I)}{E_L} \tag{11.2}$$

where P_H is the power delivered by the harvester, P_I is the power consumed by the interface during the charging time of the supercapacitor and E_L is the energy required for a single load operation, i.e., data acquisition, elaboration, and transmission.

The block diagram of the system is presented in Figure 11.13. A supercapacitor is utilized as the energy storage element instead of a battery. A 10 mF supercapacitor with the size of $12 \times 12.5 \times 2.9$ mm is selected for this harvesting system. The capacitance value of the capacitor is chosen by using Eq. (11.3), below.

$$E_C = \frac{1}{2} C_s \left(V_a^2 - V_b^2 \right) \tag{11.3}$$

where E_C is the energy needed for a single load operation, around 8mJ, V_a is assigned as 2.32V by the threshold voltage of the voltage detector, and V_b is the 1.8V, which is the minimum voltage level of the charge pump.

Besides its small size, the selected supercapacitor provides low leakage current, which is quite desirable to reduce consumed power. The power management unit is composed of a 2.32V voltage detector and a set of metal oxide field effect transistors

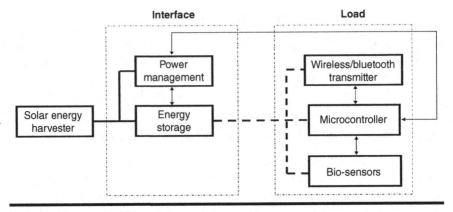

Figure 11.13   Proposed system block diagram [46].

(MOSFETs). Furthermore, the system uses a charge pump with 3.3V regulated output voltage to provide stabilized supply to the load.

The power management circuit controls the power flow from the supercapacitor to the load. When the voltage across the supercapacitor is below the threshold voltage, 2.32V, the output of the detector is low and all the MOSFETs are turned off. If the threshold voltage reaches 2.32V, the output of the detector goes high and M_1 and M_2 are turned on to supply power to the load by using the harvester and the supercapacitor. MCU generates the end-of-transmission (EOT) signal when a single load operation is finished. Then, EOT turns on M_3 and M_4, and supercapacitor charging phase restarts. Voltage detector has a small hysteresis and below 2.3V, the output of the comparator becomes low. This may become before a single load operation is performed. In order to prevent such undesired scenarios, a feedback diode is used between the drain of the M_1 and the gate of the M_2. This can keep the M_2 turned on even when the voltage level falls below 2.3V (Figure 11.14).

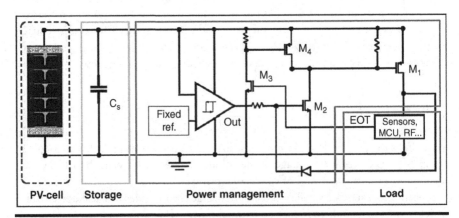

Figure 11.14   Circuit diagram of the overall system [46].

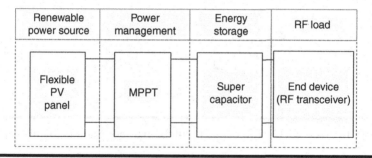

Figure 11.15   Block diagram of flexible energy harvesting system [47].

Another PV harvesting structure is presented in [47] for autonomous wearable WBANs. The block diagram is provided in Figure 11.15. In this figure, according to [46] as well, the proposed flexible energy harvesting (FEH) system uses a super-capacitor as the energy storage element. The FEH mechanism uses an ultra-low power management system (PMC) specially designed on a flexible printed circuit board (PCB). The flexible PMC has a power consumption of 32.86 μW.

Experiments for FEH system indicate that the wearable WBAN is able to monitor the temperature, read, and transmit back to the base node using wireless communication even for the indoor conditions with the typical average light intensity of 320 lux. The peak power which can be harvested from the flexible PV panel (60 mm × 72 mm) is 77 μW with 320 lux light intensity, and the FEH sensor node flexed at the angle of 30° generates 56 μW of electrical power. This means that it is possible to maintain the operation for more than 15 hours. Also, the study claims that the prototype is self-sustainable at any light intensity more than 320 lux.

Even though PV energy harvesting is quite sensitive to temperature and irradiation, there are many studies on using PV harvesters even for indoor conditions. The studies on PV energy harvesters presented in this chapter utilize supercapacitors instead of batteries. Although using supercapacitors is a much better solution than using batteries as a power supply, an ideal system is expected to work even without an energy storage element. Therefore, we can expect to see many studies in the future which focus on removing the energy storage elements and decreasing the size of the system for maintenance-free designs and mobility.

11.4.2 Thermoelectric Energy Scavenging for WBANs

Thermoelectric energy scavenging is one of the most promising harvesting technologies to be used in WBANs, since the human body can help to create a constant temperature difference between the hot and cold surfaces of TEGs. Figure 11.16 shows a TEG in contact with the human body. This type of contact can provide a constant temperature difference between the human body and the surrounding air [48]. As explained by the Seebeck effect, the constant temperature difference

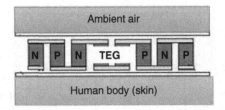

Figure 11.16 TEG surfaces in contact with human body and ambient air [48].

between the surfaces can be converted into electrical energy. Maximum efficiency of a TEG can be calculated by using Eq. (11.1). The equation clearly shows that having higher temperature differences between the surfaces increases the efficiency of the TEG. When the temperature difference is low between these surfaces, the output voltage can be very low to be used in a typical WBANs application. In such cases, the output voltage is necessary to be boosted to the desired level. Also, increasing the number of TEGs can provide a better output power.

A thermal energy-harvesting system has been proposed by [49] to power a WBAN by using human body temperature (Figure 11.17). The system has been designed to detect any falling event. An accelerometer is used to sense the fall

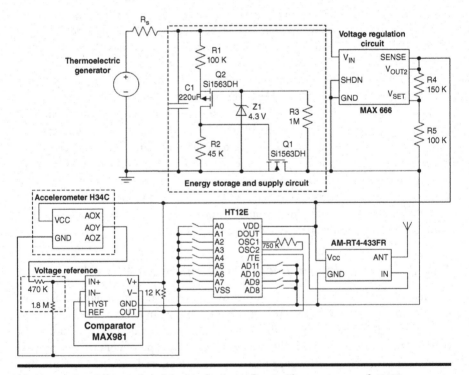

Figure 11.17 Schematic of thermal energy harvesting sensor node [49].

detection. The harvested energy is stored in a capacitor to have a higher energy level, which is necessary to power the loads. Energy storage and release events are controlled by two MOSFETs. For charging the capacitor, both MOSFETs, Q1 and Q2, are turned off to isolate the circuit from the RF load. When the voltage level across the capacitor becomes 4.9V, the preset voltage level, Q1, and Q2 are turned on so that the capacitor voltage is discharged through the voltage regulator. The voltage of 4.9V is stepped down to 3.3V by the voltage regulator and transferred to the loads for sensing and communicating operations.

Maximum electrical power generated by the TEG changes from 40 μW to 520 μW for the temperature gradient from 3°C to 15°C, respectively, for the same load resistance of 16 kΩ. However, the necessary power to drive the sensor node is reported as 14 mW. Thus, the energy storage system explained above is used to obtain the desired power level. Charging time of the capacitor is reported as less than 30 seconds with the temperature gradient of 15°C across the TEG. Experimental results of the prototype show that 1.369 mJ of energy is needed to power the loads. Also, sensed information is transmitted in 120 milliseconds in 5 digital words of 12-bit data.

Another thermal harvesting system is presented in [50] to power wearable wireless healthcare systems. The overall system architecture is shown in Figure 11.18. A bracelet-like device is introduced for powering the sensor using a TEG. The sensed data is sent to a multi-platform smartphone application for continuous monitoring. Also, the variations in voltage and power levels can be handled by the system to overcome the problems they may cause. A TEG with the size of 40 mm x 40 mm is used for the architecture. The TEG produces 20 mV with a thermal gradient of 1°C. The generated voltage with a 1°C of thermal gradient is sufficient to trigger the chip-based system. The external transformer (1:100) and DC–DC boost converter amplify the input voltage to get the higher output voltage depending on the user

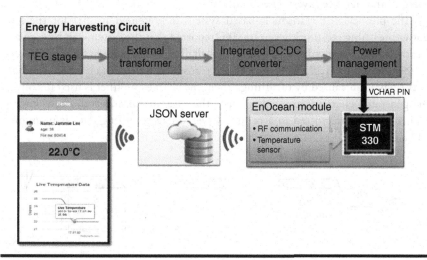

Figure 11.18   Full system architecture for [50].

Table 11.4 Measurements of End Results for the Harvester [50]

Temperature Gradient	3°C
Start-up Voltage	23.6 mV
Output Voltage	3 V
Output Power Range	50 μW to 15 mW
Output Regulation	Yes

need. The provided output voltage level can be stored in a supercapacitor and can be used as a power supply for low power devices. Table 11.4 summarizes the measurements of end results for the proposed system.

Thermal energy scavenging is quite an important method for WBANs since human body temperature can be used to create a thermal gradient. However, low output power can be a problem for low temperature gradients. Therefore, both of the presented thermoelectric harvesting studies require a voltage boost up to power WBANs. However, developments of low-power DC–DC converters are quite promising for thermoelectric energy scavenging. Thus, we can expect to see more TEG studies for WBANs in the future.

11.4.3 Piezoelectric Energy Scavenging for WBANs

Vibration or mechanical movement can be converted into electrical energy in three different mechanisms as explained in Section 11.2. In this study, piezoelectric energy harvesters are investigated in detail.

The piezoelectric transducer-based system is presented in [51] for the low-power breathing monitoring system. It is reported that the system can be used as a wearable device around the chest using a belt. The overall system architecture can be seen in Figure 11.19. The piezoelectric transducer generates charge in response to the vibration caused by breathing. Then, the charge amplifier is used to convert the generated charge into a voltage. For the ease of transmission, the voltage is digitized and transmitted to a central server where breathing data is monitored and stored.

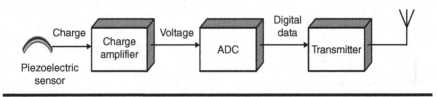

Figure 11.19 Proposed system diagram [51].

An ampulse radio ultra-wideband transmitter is used for this transmission. A 3V-600 mAh Li-Poly battery with the size of 1.6 cm × 1.6 cm is used to power the system. The battery is able to supply the system for about 40 days. However, the transducer is used as a sensor, and it does not consume any static power. Therefore, the breathing data is collected effectively without disturbing patients. Furthermore, processing and wireless transmission of the breathing data use minimal resources thanks to the customized integrated circuit design. In fact, the proposed structure uses piezoelectric transducers, not as a power supply to the system, but as a sensor. This different approach can be used in hybrid systems to make the sensing more effective where the rest of the circuit is powered by the other scavenging methods.

Piezoelectric transducers can be used not only as a sensor but also as a power supply to the system. However, they need a rectification since they provide AC power. Also, complex interface circuits and high output impedance of the piezoelectric energy harvesters are challenges for their utilization. Although there are challenges, piezoelectric energy-scavenging systems are quite attractive, since they use vibration as the input source. Vibration produced by human daily activity can be used as a source to power WBANs. Thus, the studies focus on the effective utilization of piezoelectric energy harvesting systems, and this kind of energy harvesting is promising, particularly for WBANs.

11.4.4 RF Energy Scavenging for WBANs

Despite its low output power, RF energy scavenging is an emerging technology due to the availability of the input sources. In addition, RF input sources are not affected too much by the environmental conditions like temperature and weather. Therefore, RF energy harvesting is one of the most popular research topics in recent years. Besides the availability of the input sources, utilization of RF energy scavenging may also enable wireless charging of WBANs.

The RF harvesting power management unit (PMU) is presented by [52] for the batteryless WBANs. The PMU is designed for appropriate duty-cycled operation. The duty-cycled operation is divided into two: energy charging and discharging time. The proposed PMU is able to detect both timings so that appropriate activation can be recognized. The appropriate activation provides more efficient operation and more stable wireless communication. The generic view of the system is depicted in Figure 11.20. Hysteresis comparator (H-CMP) and RF signal detector (RF-SD) components are used to detect exact timings. In active phase, the PMU generates 0.5V regulated output voltage from the charged energy.

Basically, the system starts operation by charging the storage capacitor using the RF signal. When H-CMP realizes OUT_{EH} is more than the upper threshold voltage, EN_{CMP} becomes high. Afterward, RF-SD detects the stopping of the RF signal and EN_{SD} becomes high as well. As long as both signals are high, active phase starts by activating the LDO. When voltage level across the storage element falls below the lower threshold voltage, LDO is deactivated and charging phase starts again.

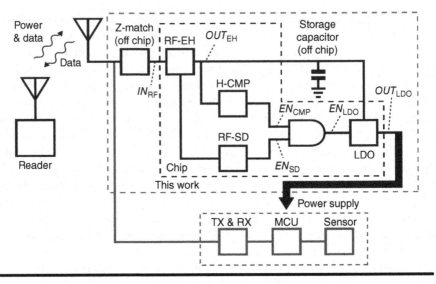

Figure 11.20   Wireless sensor system with the proposed PMU [52].

Widely available RF signals increase the importance and potential of RF energy-scavenging systems, especially for WSNs including WBANs. However, the low output voltage and power are still a problem for RF energy harvester circuits. Hence, it can be expected to see more studies on power management units, antennas, and rectifiers to increase the efficiency of RF energy harvesters. Also, multi-band RF energy-harvesting systems can be an important research area to extract more energy from RF energy-scavenging circuits and utilize RF harvesters to WBANs. Furthermore, design of wearable antennas and circuits for RF scavenging and integration possibility make this type of harvesting method very attractive for WBANs.

11.4.5 *Hybrid Energy Scavenging for WBANs*

The trend in energy harvesting is combining harvesting methods to extract energy simultaneously from multiple sources. Hybridization of energy-harvesting techniques provides multiple advantages like higher output voltage and power density compared to a single source scenario. Furthermore, using multiple sources could be an advantage for continuous operation, since in case one of the sources is not available, the other one may be used to power a WBAN [53]. Therefore, it is expected to see more studies for hybrid structures with better energy extraction efficiencies.

A hybrid energy-scavenging structure is proposed by [45] for WBAN applications (see Figure 11.21). In this study, a TEG and a solar cell are combined to extract the necessary energy. Simply, energy extracted from the TEG is used to

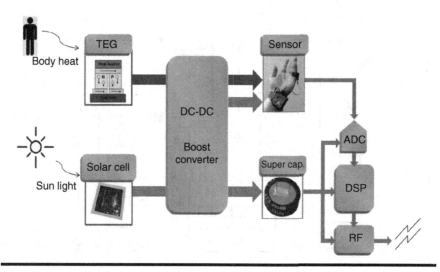

Figure 11.21   Proposed hybrid scavenging system for WBAN [45].

supply power only for the sensor. However, the extracted energy is very limited for TEGs. Thus, a solar cell is added to supply power both for the sensor and the ADC, DSP, and RF circuits. The extracted energy is used both for the sensor and charging the supercapacitor when the solar cell is activated. If the stored energy is enough, signal processing and wireless transmissions are performed. Moreover, a dual-input, dual-output DC–DC converter is presented which is able to use both of the input sources simultaneously.

The dual-input, dual-output DC–DC converter structure proposed for [45] is presented in Figure 11.22. The working principle of the converter is summarized in Figure 11.23. When only TEG is enabled, at first, S_{p1} is turned on and S_{p2} is turned off. S_n is controlled by the oscillator. When S_n is turned on, the inductor is charged, and when S_n is turned off, the stored energy is transferred to the sensor. S_n is turned on and off with a constant frequency. When both of the input sources are enabled, the system charges the inductor in the same way. However, the stored energy is transferred to both output terminals simultaneously. Table 11.5. summarizes the properties of the proposed harvester.

A hybrid energy harvesting technique for a WBAN is explained here to show an example usage of hybrid energy scavengers. However, it can be expected to see hybrid structures with better efficiency and output power, since the hybridization is one of the future trends in energy-harvesting technology. Also, hybridization is quite a promising solution to increase the capability and communication range of WBANs. Moreover, recent studies focus on combining two harvesting modes. Even though there are some studies on triple harvesting modes, further study is required for multiple energy harvesting modes.

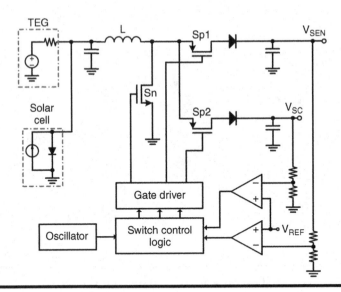

Figure 11.22 **Dual-input dual-output DC–DC boost converter for hybrid energy scavenging [45].**

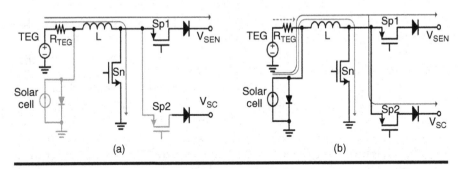

Figure 11.23 **DC–DC converter (a) TEG Mode and (b) TEG + solar cell mode operation [45].**

11.5 Conclusion and Open Issues

In this chapter, energy harvesting methods and their utilization specifically for WBANs have been investigated in detail. Energy-scavenging methods have been classified based on the ambient energy sources which they use to generate electrical energy. The classification has been performed by considering the ambient energy sources because of the fact that utilization of the harvesters depends highly on the availability of the ambient sources. The output voltages, output power densities, sizes, and ideal conditions for each harvesting mode have been mentioned.

Table 11.5 Performance Summary of the Hybrid Energy Harvester [45]

Technology	0.18 μm CMOS 1P 6M
Area	1160 μm × 1000 μm
Min. TEG Voltage	0.5 V
Min. Solar Cell Voltage	0.5 V
V_{SENSOR} / Load Current	0.9 V / 100 μA
V_{SC} / Load Current	1 V / 100 μA
Inductor / ESR	4.7 μH / 240 mΩ
Oscillator Frequency	300 kHz
Power Consumption	0.23 μW
Max. Efficiency	%55

Also, the generic view of WBANs has been presented because of the high potential of the integration of energy harvesters and WBANs to replace batteries. The batteries are bulky and invasive devices which significantly reduce mobility and create maintenance problems, especially for body area network applications. Energy harvesting is one of the best solutions to overcome these problems and improve the capabilities and communication range of WBANs. Therefore, proper design of energy-harvesting interface circuits for specific WBANs is one of the future trends. In order to design the harvesting circuits properly, application of WBANs should be carefully investigated so that output power, size, and cost constraints are going to be met at the end. In this chapter, energy-harvesting interface circuits, particularly for WBANs, are covered. Potential integration of the harvesters and WBANs has been investigated. Some studies found in the literature for different harvesting modes have been presented. Their structures, especially power management units, have been examined in detail. The sizes and corresponding output powers have been mentioned to analyze the suitability of the harvester for WBANs. It has been noticed that most of the proposed structures utilize supercapacitors as an energy storage element instead of the bulky big batteries. Although supercapacitors are smaller in size, ideally energy scavenging aims to completely remove the energy storage elements from the systems. Therefore, even though there are some studies, removing energy storage units by using harvesting structures can be still a good research topic. Also, hybridization of the harvesting modes is one of the most emerging trends in this area because of its potential in using all the available ambient resources simultaneously.

References

1. Muhtaroglu A. 2017. Micro-scale energy harvesting for batteryless information technologies. In *Energy harvesting and energy efficiency: Technology, methods, and applications*, vol. 37. N. Bizon, N.M. Tabatabaei, F. Blaabjerg, E. Kurt, eds. New York: Springer, 63–85.
2. Al-Turjman F. 2017. 5G-enabled devices and smart-spaces in social-IoT: An overview, *Elsevier Future Generation Computer Systems*, 2017. DOI: 10.1016/j.future.2017.11.035.
3. Al-Turjman F. 2017. Price-based data delivery framework for dynamic and pervasive IoT, *Elsevier Pervasive and Mobile Computing Journal*, 42, 299–316.
4. Al-Turjman F., Al-Fagih A.E., Hassanein H.S. February 2012. A novel cost-effective architecture and deployment strategy for integrated RFID and WSN systems. *International Conference On Computing, Networking And Communications (ICNC)*, 835–839.
5. Al-Turjman F., Alturjman S. 2018. Context-sensitive access in Industrial Internet of Things (IIoT) healthcare applications. *IEEE Transactions on Industrial Informatics*, 14(6), 2736–2744.
6. Movassaghi S., Abolhasan M., Lipman J., Smith D., Jamalipour A. Third Quarter 2014. Wireless body area networks: A survey. *IEEE Communications Surveys & Tutorials*, 16(3), 1658–1686.
7. Ibarra E., Antonopoulos A., Kartsakli E., Rodrigues J., and Verikoukis C. January 2016. QoS-aware energy management in body sensor nodes powered by human energy harvesting. *IEEE Sensors Journal*, 16(2), 542–549.
8. Qureshi F.U., Muhtaroğlu A., Tuncay K. April 2017. Near-optimal design of scalable energy harvester for underwater pipeline monitoring applications with consideration of impact to pipeline performance. *IEEE Sensors Journal*, 17(7), 1981–1991.
9. Doğan M., İnam S.Ç., Sürel Ö.O. 2017. Efficient energy harvesting systems for vibration and wireless sensor applications. In *Energy harvesting and energy efficiency: Technology, methods, and applications*, vol. 37, N. Bizon, N.M. Tabatabaei, F. Blaabjerg, E. Kurt. eds. New York: Springer, 87–106.
10. Cong P., Ko W., Young D. April 2010. Integrated electronic system design for an implantable wireless batteryless blood pressure sensing microsystem. *IEEE Communications Magazine*, 24(4), 98–104.
11. Harb A. 2011. Energy harvesting: State-of-the-art. *Renewable Energy*, 36, 2641–2654.
12. Rashidzadeh H., Kasargod P.S., Supon T.M., Rashidzadeh R., Ahmadi M. May 2016. Energy harvesting for IoT sensors utilizing MEMS technology. *2016 IEEE Canadian Conference on Electrical and Computer Engineering (CCECE 2016)*, 1–4, May 2016.
13. Roundy S., Wright P.K., and Rabaey J. July 2003. A study of low level vibrations as a power source for wireless sensor nodes. *Comput. Commun.*, 26, 1131–1144.
14. Lu C., Raghunathan V., Roy K. September 2011. Efficient design of micro-scale energy harvesting systems. *IEEE J. Emerging Sel. Topics Circuits Syst.*, 1(3), 254–266.
15. Veni Selvan K., Ali M.S.M. February 2016. Micro-scale energy harvesting devices: Review of methodological. *Renewable and Sustainable Energy Reviews*, 54, 1035–1047.
16. Zahid Kausar A.S.M., Reza A.W., Saleh M.U., Ramiah H. October 2014. Energizing wireless sensor networks by energy harvesting systems: Scopes, challenges and approaches. *Renewable and Sustainable Energy Reviews*, 38, 973–989.

17. KarimShaikh F., Zeadally S. March 2016. Energy harvesting in wireless sensor networks: A comprehensive review. *Renewable and Sustainable Energy Reviews*, 55, 1041–1054.
18. Al-Turjman F. 2017. Energy–aware data delivery framework for safety-oriented mobile IoT, *IEEE Sensors Journal*, 18(1), 470–478.
19. Gould C., Edwards R. September 2016. Review on micro-energy harvesting technologies. *51st International Universities Power Engineering Conference (UPEC)*, 1–5.
20. Fraas L., Partain L., 2009 *Solar cells and their applications*, 2nd ed. Hoboken, N.J.: Wiley.
21. Lu C., Park S.P., Raghunathan V., Roy K. March 2010. Efficient power conversion for ultra low voltage micro scale energy transducers. *Design, Automation & Test in Europe Conference & Exhibition* (DATE 2010), 1602–1607.
22. Muhtaroglu A. 2012. Power management and energy scavenging. In *Energy-Aware Systems and Networking for Sustainable Initiatives*, N. Kaabouch, W. C. Hu, eds. IGI Global, 310–340. North Dakota, USA
23. Brunelli D., Benini L., Moser C., Thiele L. March 2008. An efficient solar energy harvester for wireless sensor nodes. *Design, Automation and Test in Europe*, 104–109.
24. Kobayashi K., Matsuo H., Sekine Y. June 2004. A novel optimum operating point tracker of the solar cell power supply system. *IEEE 35th Annual Power Electronics Specialists Conference*, 3, 2147–2151.
25. Kim T.-Y., Ahn H.-G., Park S.K., Lee Y.-K. June 2001. A novel maximum power point tracking control for photovoltaic power system under rapidly changing solar radiation. *IEEE International Symposium on Industrial Electronics Proceedings (ISIE)*, 2, 1011–1014.
26. Senivasan S., Drieberg M., Singh B.S.M., Sebastian P., Hiung L.H. March 2017. An MPPT micro solar energy harvester for wireless sensor networks," *IEEE 13th International Colloquium on Signal Processing & its Applications (CSPA)*, 159–163.
27. Ibrahim M.A.A., Aboudina M.M., Mohieldin A.N. July 2017. An ultra-low-power MPPT architecture for photovoltaic energy harvesting systems. *IEEE EUROCON-17th International Conference on Smart Technologies*, 201–205.
28. Penella M.T., Gasulla M. May 2007. A review of commercial energy harvesters for autonomous sensors. *IEEE Instrumentation & Measurement Technology Conference (IMTC)*, 1–5.
29. Vakil A., Bajwa H. May 2014. Energy harvesting using graphene based antenna for UV spectrum. *Long Island Systems, Applications and Technology (LISAT) Conference*, 1–4.
30. Mateu L., Codrea C., Lucas N., Pollak M., Spies P. October 2007. Human body energy harvesting thermogenerator for sensing applications. *International Conference on Sensor Technologies and Applications (SENSORCOMM)*, 366–372.
31. Ramadass Y.K., Chandrakasan A.P. October 2010. A batterly-less thermoelectric energy harvesting interface circuit with 35 mV startup voltage. *IEEE Journal of Solid-State Circuits*, 46, 333–341.
32. Micropelt Inc http://www.micropelt.com/.
33. Mypelt simulation tool., www.micropelt.com/products/mypelt.php.
34. Zhang Y., Zhang F., Shakhsheer Y., Silver J.D., Klinefelter A., Nagaraju M., Boley J., Pandey J., Shrivastava A., Carlson E.J., Wood A., Calhoun B.H., Otis B.P., January 2013. A batteryless 19 W MICS/ISM-Band energy harvesting body sensor node SoC for ExG applications. *IEEE Journal of Solid-State Circuits*, 48, 199–213.

35. Wei C., Jing X. July 2017. A comprehensive review on vibration energy harvesting: Modelling and realization. *Renewable and Sustainable Energy Reviews*, 74, 1–18.

36. Amirtharajah R., Chandrakasan A.P. May 1998. Self-powered signal processing using vibration-based power generation. *IEEE Journal of Solid-State Circuits*, 33, 687–695.

37. Meninger S., Mur-Miranda J.O., Amirtharajah R., Chandrakasan A., Lang J.H. February 2001. Vibration-to-electric energy conversion," *IEEE Transactions on Very Large Scale Integration (VLSI) Systems*, 9, 64–76.

38. Hehn T., Manoli Y. 2015. Piezoelectricity and energy harvester modelling." In *CMOS Circuits for Piezoelectric Energy Harvesters*, Springer Series in Advanced Microelectronics, vol 38. Springer, Dordrecht.

39. Mhetre M.R., Nagdeo N.S., Abhyankar H.K. April 2011. Micro energy harvesting for biomedical applications: A review. *3rd International Conference on Electronics Computer Technology*, 3, 1–5.

40. Akbari S. September 2014. Energy harvesting for wireless sensor networks review. *Federated Conference on Computer Science and Information Systems*, 987–992.

41. Cao S., Li J. September 2017. A high efficiency twin coil ferrite rod antenna for RF energy harvesting in AM band. *5th International Conference on Enterprise Systems (ES)*, 276–280.

42. Uluşan H., Chamanian S., Pathirana W.P.M.R., Zorlu Ö., Muhtaroğlu A., Külah H. January 2018. A triple hybrid micropower generator with simultaneous multi-mode energy harvesting. *Smart Materials and Structures*, 27(1), 014002.1–014002.8.

43. Pantelopoulos A., Bourbakis N.G. January 2010. A survey on wearable sensor-based systems for health monitoring and prognosis. *IEEE Transactions on Systems, Man, and Cybernetics, Part C (Applications and Reviews)*, 40(1), 1–12.

44. Penderz J., van Hoof C., Gyselinckx B. 2011. Bio-Medical application of WBAN: Trends and examples. In *Bio-Medical CMOS ICs*, H.-J. Yoo, C. van Hoof, eds. New York: Springer, 279–302.

45. Wang S.-W., Im J.-P., Cho G.-H., November 2011. Dual-input dual-output energy harvesting DC-DC boost converter for wireless body area network. *IEEE Biomedical Circuits and Systems Conference (BioCAS)*, 217–220.

46. Liberale A., Dallago E., Barnabei A.L., October 2014. Energy harvesting system for wireless body sensor nodes. *IEEE Biomedical Circuits and Systems Conference (BioCAS) Proceedings*, 416–419.

47. Toh W.Y., Tan Y.K., Koh W.S., Siek L. July 2014. Autonomous wearable sensor nodes with flexible energy harvesting. *IEEE Sensors Journal*, 14(7), 2299–2306.

48. Saida M., Zaibi G., Samet M., Kachouri A. December 2016. Improvement of energy harvested from the heat of the human body. *17th International Conference on Sciences and Techniques of Automatic Control and Computer Engineering (STA)*, 132–137.

49. Hoang D.C., Tan Y.K., Chng H.B., Panda S.K. November 2009. Thermal energy harvesting from human warmth for wireless body area network in medical healthcare system., *International Conference on Power Electronics and Drive Systems (PEDS)*, 1277–1282.

50. Kanan R., Bensalem R. April 2016. Energy harvesting for wearable wireless health care systems. *IEEE Wireless Communications and Networking Conference*, 1–6.

51. Mahbub I., Wang H., Islam S.K., Pullano S.A., Fiorillo A.S. May 2016. A low power wireless breathing monitoring system using piezoelectric transducer. *IEEE International Symposium on Medical Measurements and Applications (MeMeA)*, 1–5.
52. Shirane A., Ito H., Ishihara N., Masu K. March 2015., An RF energy harvesting power management circuit for appropriate duty-cycled operation. *Japanese Journal of Applied Physics*, 54(4S), 04DE11.1–04DE11.6.
53. Al-Turjman F. 2018. QoS–aware data delivery framework for safety-inspired multimedia in integrated vehicular-IoT, *Elsevier Computer Communications Journal*, 121, 33–43.

Index

Note: Page numbers in **bold** indicate tables and those in *italics* indicate figures.

5G. *See* Fifth generation of wireless networks (5G)

A

Acetylcholinesterase (AChE), 33
AChE. *See* Acetylcholinesterase (AChE)
Adaptive WBAN. *See* Wireless body area network (WBAN)
ADC/DAC. *See* Analog-to-digital/digital-to-analog converters (ADC/DAC)
Additive white Gaussian noise (AWGN), 193
Ad-hoc on-demand distance vector (AODV), 156
Age model, 96–97, *See also* Caching approach, for WBAN
AGNR. *See* Armchair graphene nanoribbon (AGNR)
AHP. *See* Analytic hierarchy process (AHP)
ALD. *See* Atomic layer deposition (ALD)
Algorithms. *See* Routing algorithms
Amperometric sensors, 33
AN. *See* Artificial noise (AN)
Analog beamforming, 49–50, *50*
Analog-to-digital/digital-to-analog converters (ADC/DAC), 49–50
Analytic hierarchy process (AHP), 89
AN-based beamforming, 52
Angle-of arrival (AoA), 53
Antenna designs, 49–51
　　analog beamforming, 49–50, *50*
　　hybrid (analog/digital) beamforming, *50*, 50–51
　　lens-array aided beamforming, 51, *51*
Antenna subset modulation (ASM), 52–53

AoA. *See* Angle-of arrival (AoA)
AODV. *See* Ad-hoc on-demand distance vector (AODV)
Architecture, IoNT, *8*, 8–9, 133–135
　　gateways, 9, 134
　　nano-micro interface devices, 9, 133–134
　　nanonodes, 8, 133
　　nanorouters, 9, 133
Armchair graphene nanoribbon (AGNR), 13
Artificial noise (AN), 52, 53
ASM. *See* Antenna subset modulation (ASM)
Atomic layer deposition (ALD), 26
Average request per publisher (ARP), 101
AWGN. *See* Additive white Gaussian noise (AWGN)

B

Backward-learning, 152
BAN. *See* Body area network (BAN)
Base station (BS), 53
Batteryless network, 119–126, **126**
Beamforming
　　analog, 49–50, *50*
　　AN-based, 52
　　hybrid (analog/digital), *50*, 50–51
　　lens-array aided, 51, *51*
　　partial MRT (PMRT), 52
BER. *See* Bit error rate (BER)
Binary phase shift keying (BPSK), 193
Bit error rate (BER), 116
Body area network (BAN), 6, *See also* Wireless body area network (WBAN)
　　nanosenors and, 36–37
BPSK. *See* Binary phase shift keying (BPSK)
BS. *See* Base station (BS)
Buddy unicast routing, 154

233

C

Cache aware target identification (CATT),
    91, *See also* Content-oriented
    network (CON)
Cache hit ratio, 101
Caching approach, for WBAN, 88–109
    age model, 96–97
    channel communication model, 97
    content-based, 92–93
    delay model, 95–96
    ICN-based WBAN model, 94–95
    location-based, 91–92
    node functionality-based, 93–94
    popularity of on-demand requests, 97
CAGR. *See* Compound annual growth
    rate (CAGR)
Carbon nanotube (CNT), 25, 29
CATT. *See* Cache aware target
    identification (CATT)
CCN. *See* Content-centric network (CCN)
Channel-aware routing protocol, 154
Channel communication model, 97
Chemical tag sensor. *See* RFID/NFC chemical
    tag sensor
Chemical vapor deposition (CVD), 26
Cloud computing, 6
Cluster-based routing algorithms, 15
Clustering, 37
CMOS. *See* Complementary metal oxide
    semiconductor (CMOS)
CNT. *See* Carbon nanotube (CNT)
Code encoding, LCPC codes, 180–183
Cognitive-based routing algorithms, 15
Cognitive nanorouter (CNR). *See* Nano-micro
    interface devices
Cognitive node, 94, *95*
    global, 94
    local, 94
Cognitive routing protocol, 130–147
    communication model, 137–139
    energy conservation/dead node, 136–137
    IoNT architecture, 133–135
    lifetime in IoNT, 135–136
    overview, 130–133
    simulation results, 143–147
Communication
    molecular, 13
    nano-electromagnetic, 13
Communication model, 137–139
Communication protocols, 13–15
    MAC, 14

routing algorithms, 14–15
wireless communication models, 13–14
Communication technologies, comparison
    of, **12, 158**
Complementary metal oxide semiconductor
    (CMOS), 46
Complexity analysis, LCPC codes, 191–193
Compound annual growth rate (CAGR), 24
CON. *See* Content-oriented network (CON)
Confidentiality, millimeter (mm)-waves, 51–53
Connectivity as design factor, 11
Constraints of IoNT, 15–16
    energy consumption as, 15
    primary, 15
    secondary, 15
Content-based caching, 92–93
Content-centric network (CCN), 93
Content-oriented network (CON), 91
Continuous aperture phased (CAP)-MIMO.
    See Lens-array aided beamforming
Converters
    ADC/DAC, 49–50
    DC-DC, 225, *226*
Cooperative, WBASN, 115, *115*, 117, 118
Coordinate-based addressing scheme
    (CORONA), 155
CORONA. *See* Coordinate-based addressing
    scheme (CORONA)
Cost as design factor, 11
Cryptography, 51
Current. *See* Direct current (DC)
CVD. *See* Chemical vapor
    deposition (CVD)
Cyclic voltammetry, 33

D

Data multiplexing, 51
Data-oriented network architecture
    (DONA), 91
Data traffic, 66–67
DC. *See* Direct current (DC)
DC–DC converter, 225, *226*
Decoding, LCPC codes
    error correction, 188–189, *189*
    error detection, 184–188, *189*
Defect dynamics, 23, *See also* Nanostructured
    materials
Delay as design factor, 11
Delay model, WBAN, 95–96
Delay time, 143
Denial of service (DoS), 51

Design, 9–12, **12**
  antenna, *See* Antenna designs
  connectivity, 11
  cost, 11
  delay, 11
  energy harvesting, *See* Energy harvesting
  security, 10
  short wavelength, 9–10
Destination sequenced distance vector
  (DSDV), 156
DIF. *See* Dynamic infrastructure (DIF)
Digital signal processor (DSP), 208
Direct current (DC), 207
DONA. *See* Data-oriented network
  architecture (DONA)
DoS. *See* Denial of service (DoS)
Drug delivery systems, *28*, 28–29
DSDV. *See* Destination sequenced distance
  vector (DSDV)
DSP. *See* Digital signal processor (DSP)
Duty cycling, 37
Dynamic infrastructure (DIF), 155

E

E3A. *See* Enhanced energy-efficient
  approach (E3A)
Eavesdropping, 51
ECG. *See* Electrocardiogram (ECG)
Electrical energy, 205
Electrocardiogram (ECG), 25
Electrochemical sensors, 33–35
  amperometric, *33*
  potentiometric, 34–35
  voltammetric, 33–34
Electromagnetic field (EMF), 16
Electromagnetic waves
  millimeter, 45–54
  terahertz, 45–46
EMF. *See* Electromagnetic
  field (EMF)
Encryption, 51, 52
End-of-transmission (EOT), 218
Energy-aware routing protocol, 151–173
  LaGOON protocol, 159–165
  overview, 151–156
  performance evaluation, 165–166
  simulation results, 167–172
Energy conservation
  data-driven, 37
  and dead node, cognitive routing protocol,
    WBAN, 136–137

duty-cycling, 37
  mobility-based schemes, 37
Energy consumption, as constraint
  of IoNT, 15
Energy efficiency .
  comparison with respect to
    payload, *123*
  for cooperative WSBAN, 117
  of one-hop with respect to payload, *123*
  payload simulations *vs.*, 122, *123*
  with respect to distance, 124
  for single hop WSBAN, 117
  for two-hop WSBAN, 118
Energy harvesting, 10, 204–227, *215*, *See also*
  Flexible energy harvesting (FEH)
  for batteryless WBASN, 119
  as design factor, 10
  nanosensor and, 35–36, 39
Energy scavenging
  hybrid, 213–214, 224–225, *225*, *226*
  microscale, 205–214
  photovoltaic, *207*, 207–208, 217–219
  piezoelectric, 222–223
  radio Frequency (RF), 212–213, **213**,
    213, 223–224
  thermoelectric, *See* Thermoelectric energy
    scavenging
  vibration-based, 210–211
Energy transfer. *See* Fluorescence resonance
  energy transfer (FRET)
Enhanced energy-efficient approach
  (E3A), 139–140
  performance evaluation of, 140–147
  performance metrics, 142–143
  performance parameters, 143
Environment monitoring and research, 38
EOT. *See* End-of-transmission (EOT)
Error correction, decoding, LCPC codes,
  188–189, *189*
Error detection, decoding, LCPC codes,
  184–188, *189*
Experimental setup
  enhanced energy-efficient approach
    (E3A), 141–142
  RDDA, 75

F

Fabrication of nanosensors, 29–35
Failure rate, 142
Fairness, 76
Fast Fourier transform (FFT), 53

FEH. *See* Flexible energy harvesting (FEH)
Feynman, Richard, 22
FFT. *See* Fast Fourier transform (FFT)
FIFO. *See* First in first out (FIFO)
Fifth generation of wireless networks (5G), 6, 53
    architecture in, 8–9
    communication protocols and, 13–15
    energy harvesting in, 10
    market opportunity in, 7–8
    physical layer communication and, **12**, 12–13
First generation of nanomaterials, 28
First in first out (FIFO), 90
Flat power spectral densities, 14, 48, 137
Flexible energy harvesting (FEH), 219, *219*
Flooding routing, 152
Fluorescein, 30
Fluorescence resonance energy transfer (FRET), 31
Food packing research, 38
Fourth generation of nanomaterials, 28
FRET. *See* Fluorescence resonance energy transfer (FRET)
Friis' Law, 47–48

G

GAF. *See* Geographic adaptive fidelity (GAF)
Gateways, 9, 36, 134
Gaussian pulse, power spectral densities, 14, 48, 137
GCN. *See* Global cognitive node (GCN)
GCN request time (seconds), 143
GEAR. *See* Geographic and energy-aware (GEAR)
Geographic adaptive fidelity (GAF), 132
Geographic and energy-aware (GEAR), 132
Global cognitive node (GCN), 94
Global positioning systems (GPS), 24
GPS. *See* Global positioning systems (GPS)
GPSR. *See* Greedy perimeter stateless routing (GPSR)
Grain boundaries, 23, *See also* Nanostructured materials
Grains, 23, *See also* Nanostructured materials
Greedy perimeter stateless routing (GPSR), 154
GW request time (seconds), 77

H

Healthcare applications, IoNT in. *See* Medicine/healthcare, IoNT in
Hub nodes. *See* Sink nodes
Hybrid (analog/digital) beamforming, *50*, 50–51
Hybrid energy scavenging, 213–214, *See also* Energy scavenging
    for WBAN, 224–225, *225*, *226*

I

ICN. *See* Information-centric network (ICN)
ICN-based WBAN model, 94–95
ICNIRP. *See* International Commission on Non-Ionizing Radiation Protection (ICNIRP)
In-body communication, 6
Information-centric network (ICN), 89
In-network latency (delay), 101
International Commission on Non-Ionizing Radiation Protection (ICNIRP), 16
Internet of Nano-Things (IoNT), *See also* Wireless body area network (WBAN)
    architecture, *See* Architecture, IoNT
    communication protocols, 13–15
    constraints of, 15–16
    defined, 6
    deployment of, 15–16
    in 5G era, *See* Fifth generation of wireless networks (5G)
    medicine/healthcare and, 24
    millimeter waves, *See* Millimeter waves
    nanosensors for, *See* Nanosensor(s)
    open research issues, 17
Internet of Things (IoT), 6, 130, 151, 204
IoNT. *See* Internet of Nano-Things (IoNT)
IoT. *See* Internet of things (IoT)

L

LaGOON (last good neighbor), 152, 159–165, *See also* Routing protocol(s)
Latency, 76
LCN. *See* Local cognitive node (LCN)
LCPC. *See* Low-complexity parity check (LCPC) codes

Learning
  enhanced energy-efficient approach (E3A),
    140
  rational data delivery approach (RDDA),
    68–69
Least recently used (LRU), 90
Least value first (LVF), 90
Lens-array aided beamforming, 51, *51*
Lifetime in IoNT, 64–65, 135–136
Line of sight (LoS), 6
Local cognitive node (LCN), 94
Location-based caching, 91–92
LoS. *See* Line of sight (LoS)
Low-complexity ASM. *See* Silent antenna
    hopping (SAH)
Low-complexity parity check (LCPC) codes,
    177–199
  code encoding, 180–183
  decoding, 184–190
  error correction, 188–189, *189*
  error detection, 184–188, *189*
LRU. *See* Least recently used (LRU)
LVF. *See* Least value first (LVF)

M

MAC protocols, 14
Market opportunity in the 5G era, 7–8
Material science and nanosensor fabrication,
    29–35
Maximum power point tracking
    (MPPT), 208
Maximum ratio transmission (MRT), 52, 53
Mechanical nanosensor, 35
Medicine/healthcare, IoNT in, 23–29
  engineered nanodevices/nanostructures, 28
  patient's body, 27–29, *28*
  patient's surrounding environment, 24–27,
    25, 26, 27
MEMS. *See* Microelectromechanical systems
    (MEMS)
Metal oxide field effect transistors (MOSFET),
    217–218
Microelectromechanical systems (MEMS), 210
Microscale energy scavenging, 205–214, *See also*
    Energy scavenging
  hybrid energy scavenging, 213–214,
    224–225, *225, 226*
  photovoltaic, *207*, 207–208, 217–219
  radio frequency (RF), 212–213, **213**, *213*,
    223–224
  techniques, *205*

  thermoelectric, 208–209, 219–222, *220*
  vibration-based, 210–211
Millimeter electromagnetic waves, 45
Millimeter waves, 45–54
  confidentiality in, 51–53
  overview, 45–47
  propagation model, 47–49
  research in, 54
MIMO. *See* Multiple input multiple
    output (MIMO)
Mm-Wave. *See* Millimeter waves
Modulation. *See* Antenna subset
    modulation (ASM)
Molecular communication, 13, 36
MOSFET. *See* Metal oxide field effect
    transistors (MOSFET)
MPPT. *See* Maximum power point
    tracking (MPPT)
MRT. *See* Maximum ratio transmission (MRT)
Multicast routing protocols, 152
Multipath routing protocols, WSN, 132
Multiple input multiple output (MIMO),
    13, 48, 49
Multiplexing. *See* Data multiplexing

N

Nanocarrier platforms, *28*, 29
  for drug delivery systems, *28*
Nanocluster composition algorithm (NCCA),
    156
Nanocommunication, 36
Nano-electromagnetic communication, 13
Nanomaterials. *See* Nanostructured materials
Nanomedicine, 27
Nano-micro interface devices, 9, 36, 133–134
Nanonetworks, 158, *159*
Nanonodes, 8, 36, 133
Nanorouters, 9, 36, 133
Nanosensor(s), 21–39, 64, *See also* Optical
    nanosensors
  body area network (BAN) and, 36–37
  defined, 22
  energy harvesting and, 35–36
  fabrication and material science, 29–35
  milestone chart of, *29*
  overview, 22
Nanosensor networks, 11
  energy-aware routing protocol, 151–173
Nanostructured materials, 23, *23*
  first generation of, 28
  fourth generation of, 28

second generation of, 28
third generation of, 28
Nanotechnology, 22, 29
NCCA. *See* Nanocluster composition
algorithm (NCCA)
Nearest neighbor algorithm (NNA), 62
enhanced energy-efficient approach
(E3A), 141
of RDDA, 74
Near-field communication (NFC), 24, 26
Near-infrared light (NIR), 30
Network. *See specific network*
Network architecture, *See also* Data-oriented
network architecture (DONA)
rational routing protocol for WBAN,
63–64, *64*
Network lifetime, 76, 142
NFC. *See* Near-field communication (NFC)
NIR. *See* Near-infrared light (NIR)
NLoS. *See* Non-line of sight (NLoS)
NLOS. *See* Non-line of sight (NLOS)
NNA. *See* Nearest neighbor algorithm (NNA)
Node functionality-based caching, 93–94
Nodes, 6, 15–16, *See also specific node*
Non-line of sight (NLoS), 6, 49, 120

O

OADA. *See* Observe analyze-decide-act
(OADA)
Observe analyze-decide-act (OADA), 134
On-body communication, 6
Optical nanosensors, 30–32
Optimal power spectral densities, 14, 48, 137

P

Packet error rate
with respect to node distance, *122*
with respect to payload, 120–122
Packet-size optimization for batteryless
WBASN, 119
Partial MRT-based (PMRT) beamforming, 52
Path length (hops), 76
PDMS. *See* Polydimethylsiloxane (PDMS)
Peer-to-peer type routing protocol, 155
Peltier effect, TEG, 208
Performance evaluation
energy-aware routing protocol for
nanosensor networks, 165–166
enhanced energy-efficient approach (E3A),
140–147

of RDDA, 73–76
VoI cache replacement, 100–108
Performance metrics
energy-aware routing protocol for
nanosensor networks, 166–167
enhanced energy-efficient approach (E3A),
142–143
rational data delivery approach (RDDA), 76
VoI cache replacement, 101
Performance parameters
enhanced energy-efficient approach
(E3A), 143
rational data delivery approach (RDDA), 77
PET. *See* Polyethylene terephthalate (PET)
Phase shifters, 49–51
Photovoltaic energy scavenging, *207*, 207–208
for WBAN, 217–219
Physical layer communication, **12**, 12–13
Physical layer security (PLS), 51–52, 53
PI. *See* Polyimide (PI)
Piezoelectric energy scavenging, for WBAN,
222–223
Piezoelectric transducers (PZT), 211, *211*
Plastic vapor deposition (PVD), 26
PLF. *See* Probability of link failure (PLF)
PLS. *See* Physical layer security (PLS)
PMC. *See* Power management system (PMC)
PMMA. *See* Polymethyl methacrylate (PMMA)
PMRT. *See* Partial MRT-based (PMRT)
beamforming
PMU. *See* Power management unit (PMU)
PNF. *See* Probability of node failure (PNF)
Polyacrylamide nanoparticle, 30
Polydimethylsiloxane (PDMS), 26
Polyethylene terephthalate (PET), 26
Polyimide (PI), 26
Polymethyl methacrylate (PMMA), 26
Popularity of on-demand requests, WBAN, 97
Potentiometric sensors, 34–35
Power management system (PMC), 219
Power management unit (PMU), 223
Power spectral densities, 14, 48, 137
flat, 14, 48, 137
Gaussian pulse, 14, 48, 137
optimal, 14, 48, 137
transmission window at 350 GHz,
14, 48, 137
Primary constraints, of IoNT, 15
Probability of link failure (PLF), 88
Probability of node failure (PNF), 88
Propagation model, millimeter
(mm)-wave, 47–49

PVD. *See* Plastic vapor deposition (PVD)
PZT. *See* Piezoelectric transducers (PZT)

Q

QoI. *See* Quality of information (QoI)
QoS. *See* Quality of service (QoS)
Quality of information (QoI), 14, 61, **89**
Quality of service (QoS), 89

R

Radio frequency identification (RFID),
    24, 26
Radio frequency integrated circuit (RFIC), 46
Radio frequency (RF), 25, 49
Radio frequency (RF) energy scavenging,
    212–213, **213**, *213*
    for WBAN, 223–224
Random point-to-point routing protocols, 61
Random routing, 152–153
Rational data delivery approach (RDDA),
    67–83
    learning, 68–69
    performance evaluation, 73–81
    reasoning, 70–73
    simulation results, 77–83
Rational routing protocol, 60–83, *See also*
    Cognitive routing protocol
    communication model, 65–66
    energy conservation, 65
    lifetime in IoNT, 64–65
    network architecture, 63–64, *64*
    overview, 60–63
    random point-to-point, 61
    simple flooding, 61
    traffic model, 66–67
RDDA. *See* Rational data delivery approach
    (RDDA)
Reasoning
    enhanced energy-efficient approach
      (E3A), 140
    rational data delivery approach
      (RDDA), 70–73
ReInForM. *See* Reliable information forwarding
    using multiple paths (ReInForM)
Relay node (RN), 94
Reliable information forwarding using multiple
    paths (ReInForM), 132
Remaining energy (joule), 76
Remaining energy (pJ), 142
Repair time, 76

Request time, 77, 142
Research, 17
    defense, 39
    energy harvesting, 39
    environment monitoring, 38
    food packing, 38
    in healthcare applications, 23–29
    in millimeter waves, 54
    traffic accidents prevention, 38–39
    viruses and bacteria, detection of, 38
RF. *See* Radio frequency (RF)
RFIC. *See* Radio frequency integrated
    circuit (RFIC)
RFID. *See* Radio frequency identification
    (RFID)
RFID/NFC chemical tag sensor, 26
RN. *See* Relay node (RN)
Routing algorithms, 14–15, 60
    cluster-based, 15
    cognitive-based, 15
Routing protocol(s)
    channel-aware, 154
    cognitive, *See* Cognitive routing protocol
    energy-aware, 151–173
    multicast, 152
    multipath, 132
    peer-to-peer type, 155
    random point-to-point, 61
    rational, *See* Rational routing protocol
    simple flooding, 61
    single-path, WSN, 132

S

SAH. *See* Silent antenna hopping (SAH)
SAR. *See* Sequential assignment routing (SAR)
Scavenging. *See* Energy scavenging
Secondary constraints, of IoNT, 15
Second generation of nanomaterials, 28
Security
    confidentiality, 51–53
    as design factor, 10
    physical layer, 51–52
Seebeck effect, TEG, 208
Sensor network. *See* Wireless body area sensor
    network (WBASN); Wireless sensor
    network (WSN)
Sensor nodes (SN), 94, *95*
Sensors, 22, *See also* Nanosensor(s)
    amperometric, 33
    potentiometric, 34–35
    voltammetric, 33–34

Sequential assignment routing (SAR), 132
Shortest path algorithm (SPA), 132
  of enhanced energy-efficient approach
    (E3A), 141
  of rational data delivery approach
    (RDDA), 73
Short wavelength as design factor, 9–10
Signal processor. *See* Digital signal
    processor (DSP)
Signal to noise (SNR), 116
Silent antenna hopping (SAH), 53
Simple flooding routing protocols, 61
Simulation parameters, of VoI cache
    replacement, 101–102
Simulation results
  energy-aware routing protocol,
    167–172
  LCPC codes, 193–199
  RDDA, 77–83
  VoI cache replacement, 102–109
Single-hop, WBASN, 115, *115*, 117
Single-path routing protocols, WSN, 132
Sink nodes, 6, 94
Sitti, Metin, 23–24
SN. *See* Sensor nodes (SN)
SNR. *See* Signal to noise (SNR)
SoC. *See* System on chip (SoC)
SPA. *See* Switched phased-array (SPA)
Spectral density. *See* Optimal power
    spectral densities
SPP. *See* Surface plasmon polariton (SPP)
Success rate, 142
Surface plasmon polariton (SPP), 13
Switched phased-array (SPA), 53
System on chip (SoC), 25

T

Tag sensor. *See* RFID/NFC chemical
    tag sensor
TDD. *See* Time division duplexing (TDD)
TEG. *See* Thermoelectric generator (TEG)
Terahertz (THz) band communication, 6, 13
Terahertz waves, 45–46, *See also*
    Electromagnetic waves
Theoretical delay analysis, VoI cache
    replacement, WBAN, 99–100
Thermoelectric energy scavenging, 208–209
  for WBAN, 219–222, *220*
Thermoelectric generator (TEG), 208–209,
    209, 221
  Peltier effect, 208

Seebeck effect, 208
  Thomson effect, 208
Third generation of nanomaterials, 28
Thomson effect, TEG, 208
Threats, 76
Time division duplexing (TDD), 52
Time-to-hit data (TTH), 101
Total number of transmission rounds, 143
Traffic accidents and research, 38–39
Traffic model, 66–67
Transmission rounds, 77
Transmission window at 350 GHz, power
    spectral densities, 14, 48, 137
TTH. *See* Time-to-hit data (TTH)
Two-hop, WBASN, 115, *115*, 118

U

UCNP. *See* Upconversion nanoparticle
    (UCNP)
Upconversion nanoparticle (UCNP),
    30–31, *32*
Utra-wide band (UWB) communication, 24

V

Value-based caching. *See* Caching approach,
    for WBAN
Value of sensed information (VoI), 89, *See also*
    VoI cache replacement
Vibration-based energy scavenging, 210–211
Viruses and bacteria, detection of, 38
VoI. *See* Value of sensed information (VoI)
VoI cache replacement, 98–100
  performance evaluation, 100–108
  performance metrics, 101
  theoretical delay analysis, 99–100
Voltammetric sensors, 33–34
Voltammetry. *See* Cyclic voltammetry

W

Waves. *See* Electromagnetic waves
WBAN. *See* Wireless body area network
    (WBAN)
WBSAN. *See* Wireless body area sensor
    network (WBASN)
WHO. *See* World Health Organization (WHO)
Wireless body area network (WBAN), 36–37,
    214–216, *216*
  adaptive, 113–125
  caching, *See* Caching approach, for WBAN

energy-harvesting methods, 204–227,
　　See also Energy scavenging
gateway, 36
LCPC codes, 177–199
nano-micro interface, 36
nanonodes, 36
nanorouters, 36
routing protocol, *See* Routing protocol(s)
Wireless body area sensor network
　　(WBASN), 113–126

cooperative, 115, *115*, 117
overview, 113–115
single-hop, 115, *115*, 117
two-hop, 115, *115*
Wireless communication models, 13–14
Wireless sensor network (WSN), 89, 90
　　multipath routing protocols, 132
　　single-path routing protocols, 132
World Health Organization (WHO), 16
WSN. *See* Wireless sensor network (WSN)

Printed in the United States
by Baker & Taylor Publisher Services

Printed in the United States
by Baker & Taylor Publisher Services